# Orientation of Single Crystals by Back-Reflection Laue Pattern Simulation

## C. Marín & E. Diéguez

*Universidad Autónoma de Madrid*

**World Scientific**
*Singapore • New Jersey • London • Hong Kong*

*Published by*

World Scientific Publishing Co. Pte. Ltd.

P O Box 128, Farrer Road, Singapore 912805

*USA office:* Suite 1B, 1060 Main Street, River Edge, NJ 07661

*UK office:* 57 Shelton Street, Covent Garden, London WC2H 9HE

**Library of Congress Cataloging-in-Publication Data**
Marín. C. (Carlos), 1970–
    Orientation of single crystals by back-reflection Laue pattern
simulation / C. Marín & E. Diéguez.
        p.    cm.
    Includes bibliographical references and index.
    ISBN 9810228716
    1. X-ray crystallography.    I. Diéguez. E. (Ernesto)    II. Title.
QD945.M247    1999
548'.83--dc21                                                    98-54444
                                                                        CIP

**British Library Cataloguing-in-Publication Data**
A catalogue record for this book is available from the British Library.

Printed in Singapore.

*To Mª Victoria and Antonio*

# PREFACE

It is well known that the Laue method was the first operative technique to use X-rays as a tool for crystal analysis. The method, as we all know, was developed in the epoch-making experiment conducted by W. Friedrich and Paul Knipping following the ideas of Max Theodor Felix von Laue some 85 years ago. It is remarkable that the original technique has been changed only in small details and remains very much like the one developed during the initial experiment. Some 20 years ago we stated that those interested in classical crystallography will discover that older crystallography acquires new significance in terms of the interpretation of crystal data available in the Laue method, which provides a speedy and simultaneous goniometry of all crystal planes of interest. We also said that this easy goniometry and the abundant information that Laue photographs contain may encourage many metallurgists (or crystal growth scientists), who ordinarily make only limited use of the information available, to take advantage of the additional information which these photographs can still provide. It was therefore a happy surprise for me to learn that E. Diéguez and C. Marín of the Department of Physics of Materials of the Universidad Autónoma de Madrid have prepared a book that deals with a specific application of the Laue method. I was very happy to realize that the old method — that has proven, in our hands, to be such a useful technique in the study of X-ray diffuse scattering of crystals — is still very much alive.

This book is the result of the hybridization of an old X-ray diffraction technique and modern computer programming and simulation techniques. The authors have much experience in the field of crystal growth and crystal orientation determination by back-reflection Laue patterns. Using this experience they have developed an algorithm that allows a straightforward method for determining the orientation of a crystal using the information gathered from a back-reflection Laue diagram. The subject matter of the book falls naturally into two parts. The first part provides an elementary survey of the fundamental knowledge of lattices and X-ray diffraction by crystals. The second part deals with the development of the algorithm, the presentation of the corresponding software and its application to crystal orientation determination using back-reflection Laue patterns. The computer software allows the simulation of back-reflection Laue grams of crystals and indexing of the Laue spots; a users' manual provides an easy to follow step-by-step routine

vii

for entering the needed data and obtaining the corresponding Laue gram simulation. A list of selected bibliography is given at the end of the book. The straightforward language used in the book by the authors makes it an easy read for specialists and students alike.

I am sure that this book will not only be a valuable complement to university curricula in materials science involving courses in crystal growth and X-ray diffraction, but also an aid to researchers involved in the study of crystals.

J. L. Amorós

# NOTE FROM AUTHORS

From the time Prof. Nicolás Cabrera began with the science of crystal growth in our university 25 years ago, the unavoidable first step for the orientation of single crystals has been a time consuming task. The increase in computer capabilities runs parallel to the decrease of their price during this period. Therefore, we have been developing algorithms to aid in orientation work in order to reduce time and costs in our daily research labour.

Many of our collaborators have encouraged us to give a 'friendly face' to our algorithms and to offer it to the PC users in the scientific community, given the current power of these computers. In this atmosphere, this book comes about as the mature fruit of all our experiences in the orientation of crystals by employing the X-ray diffraction in the back-reflection Laue geometry. Here we would like to acknowledge the support of our editors, Dr. R. Duhlev, Mr. M.Z. Chow and Ms. J.M. Tan.

After a brief description in the first chapter of the X-ray diffraction fundamentals, the second chapter is a deep study of the factors which influence the back-reflection Laue patterns, which are the support for the algorithms fully developed in the third chapter. In the fourth chapter, a complete description about the use of the attached program is given, and the fifth chapter presents several examples of experimental Laue-grams for known materials of each of the seven crystal systems.

We must point out that the main objective of the book is to allow simulation and indexing of any kind of single crystals, and in this way the program WSLAUE enclosed with the book contains the real tool needed for the orientation of any crystal.

It was in our mind to write a book using a language that can be understood by researchers working in any scientific field who, at anytime, are involved with work on the orientation of single crystals, just with the basic graduate knowledge.

It is extremely important for us to recognize that this project would be impossible without the contribution of many people both from within and outside of our Department of Physics of Materials. We specially acknowledge Dr. P.S. Dutta for his ideas, discussions and careful revision and correction of the manuscript. The collection of the samples shown in the text has been possible thanks to the contributions of Dr. Cheng Changkang, Dr. C. Zaldo, Dr. T. Duffar and Dr. J. Przeslawski. From a theoretical point of view, we have received important support from Profs. L. Arizmendi and J. Tornero, and from experimental realization one from A. Cintas and R. Fernández. The testing of the program, discussions and the encouragement from people involved in our Crystal Growth Lab during the writing of this book has been a day-to-day incentive and the main source of test samples. Therefore, we must acknowledge V. Bermúdez, M.T. Santos, and Drs. N.V. Sochinskii, M.D. Serrano and especially J.C. Rojo, to whom we are indebted for his help during the computing work as well. We must acknowledge Drs. W. Kraus and G. Nolze for allowing us to use their software PowderCell for the visualization of the crystal structures shown in chapter five, and for his interest in our project. Finally, our gratitude goes to Prof. J.L. Amorós who kindly agreed to write the preface for this book. We are honored to obtain contributions and comments from the professor who has earned the respect and admiration of many scientists. This acknowledgement must be extended to his wife Prof. M.L. Canut-Amorós, a pioneer of computer physics applied to crystallography, for her kind interest in our work.

Finally, we are indebted, for the financial support received, to the Vicerrectorado de Investigación of the Universidad Autónoma de Madrid and the Spanish Scientific Agency C.I.C.yT. under the Project TXT96-1688, and to Polaroid España S.A. and Kodak S.A. for the supply of films.

C. Marín                    E. Diéguez

Madrid (Summer 1997)

# CONTENTS

# CHAPTER 1

# X-RAY DIFFRACTION BY CRYSTAL LATTICES

The X-rays are electromagnetic radiations with wavelength ($\lambda=c/\omega$) of the order of angstroms (Å). The spectrum of the X-rays is situated between the gamma rays and the ultraviolet (UV) rays as depicted in figure 1.1. Two zones are traditionally defined for X-rays depending on their wavelength: hard and soft X-rays. Hard X-rays have a front with gamma rays at 1-0.01 Å, while the transition between the soft and the UV rays is located between 10 and 100 Å. Since the interatomic distances in solids are of the order of Å, the wavelength corresponding to X-rays is suitable for carrying out diffraction experiments. In the present chapter, we discuss the techniques of generation and detection of X-rays followed the fundamental aspects of crystal geometry and the scattering of X-rays from crystalline material.

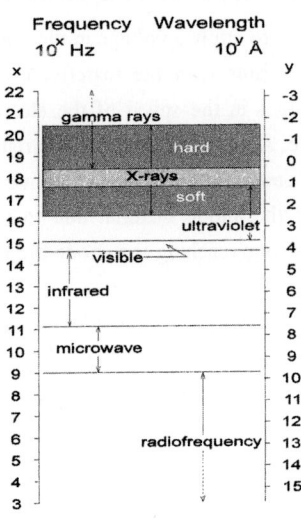

**Figure 1.1:** *Diagram of the electromagnetic spectrum.*

1

## 1.1 Production of X-rays

Electromagnetic radiation is formed by oscillatory electric and magnetic fields perpendicular to each other. For purely monochromatic waves, fields depend on time and position as,

$$\mathbf{E(r,t)} = \mathbf{E}_\omega e^{-i\omega t} \qquad \mathbf{B(r,t)} = \mathbf{B}_\omega e^{-i\omega t}. \qquad [1]$$

The intensity of electromagnetic radiation is taken as proportional to the square of electric field, as the contribution of magnetic field is negligible.

Radiation is produced due to the deceleration of a charged particle. The energy of the emitted wave being equal to the variation of the kinetic energy of the particle. When the charged particle is an electron and the speed lost is around one third of the speed of light (c), the energy of the emitted wave is in the range of the X-ray. Therefore, one way to produce X-rays is by applying a potential difference between the electrodes of a tube in such a way that an electron removed from the cathode will be accelerated by the applied potential and are stopped by the anode. If V is the applied potential and e and $m_e$ being the charge and mass of the electron respectively, the speed v is obtained from $eV = \frac{1}{2}mv^2$. Since quantum energy $E=h\nu=hc/\lambda$, for the wavelength of emission to be of the order of Å, it is necessary to apply a voltage in the range of kV.

The impact of the electrons with the material used as anode are inelastic in nature. Hence, there are changes in the speed of the electrons due to the interactions with the nucleus used as anode. As a consequence, the variation of energy of this process produces either a radiative emission or a non radiative phenomena. Non radiative processes produce heating of the anode material due to the atomic oscillations. Only about one per cent of the electrons reaching the anode are effective in the production of X-rays.

An cathodic tube X-ray spectrum can be either a continuous or a line spectrum. Figure 1.2 shows the spectral distribution of the emitted X-rays for a Mo cathode tube for different voltages applied between anode and cathode.

The continuous spectrum (also called white radiation or *bremsstrahlung*) is due to the sudden reduction of energy of the bombarding electrons. The spectrum begins abruptly at a short wavelength limit (SWL), there is an intensity maximum at a wavelength related to anode thickness, which becomes higher and shifts to shorter wavelengths as the voltage applied to the tube (V) increases. It should be noted that the wavelengths as the voltage applied to the tube (V) increases. It should be noted that the distribution continuous emission depends on the material used as anode. An empirical relationship has been formulated (1) to describe the intensity emission distribution for any material used as anode depending on the applied voltage.

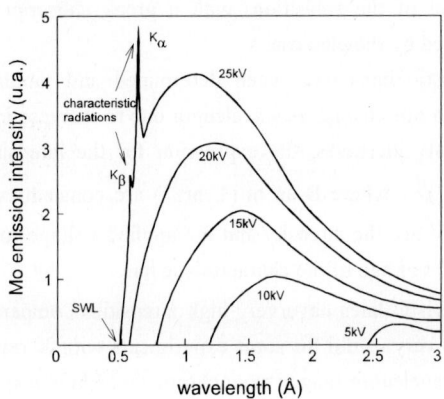

**Figure 1.2:** *Spectral emission distribution of a Mo cathodic tube for different applied voltages.*

The short wavelength limit, $\lambda_{min}$, correspond to the case where the bombarding electrons emit all their kinetic energy. Therefore, $\lambda_{min}$ is related to the voltage (V) across the X-ray tube by means of the quantum formula $eV = h\nu_{max}$, being $\lambda_{min} = c / \nu_{max}$, we have $\lambda_{min} = hc / eV$.

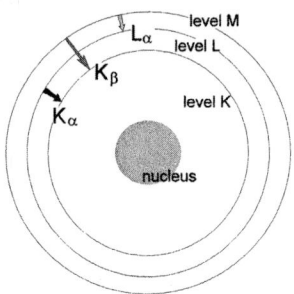

**Figure 1.3:** *Scheme of the electronic transitions in an atom.*

There is a threshold voltage, for which at a fixed wavelength the emitted intensity has an abrupt increase, as seen in figure 1.2 at 20.01 kV. These peaks shown at 25kV are named "characteristic lines" because they are specific to the anode material. Traditionally these lines are designated as $K_\alpha$, $K_\beta$, $L_\alpha$ ... which are electronic transitions of the atom as shown in figure 1.3. The nomenclature of the lines has a capital letter which signifies the final level of the transition, with a greek subscript which indicates the number of levels crossed by the electron.

The characteristic lines have been determined and tabulated (2.a) for every element. These lines do not change its wavelength due to increase in the applied voltage, nevertheless its intensity increases, the expression for the intensity of the K lines is, $I_{Kline} = Bi(V - V_{Kline})^m$, where B an m $(1 \le m \le 2)$ are constants which depend on the anode material, i and V are the intensity and the applied voltage to the source tube and $V_{Kline}$ is the threshold voltage of the characteristic line.

These characteristic lines have very high intensities compared to the continuous emission, and they are very useful for some experiments with X-rays when one needs to work with a narrow wavelength range (monochromatic techniques). However, this is not the case for obtaining a Laue-gram of a single crystal. Hence, one must work with a continuous and uniform wavelength distribution in order to get many constructive diffraction conditions.

Amongst non-conventional X-ray sources, synchrotron radiation is more frequently used as a tool in various scientific research purposes (3). The main peculiarities of synchrotron radiation obey the relativistic nature of the generation process (4), and could be summarised by the following points: high intensity and collimation of the emitted radiation, option for an intense tuneable emission, and a large spectral range which goes from the far UV to the gamma radiation.

## 1.2 Detection of X-rays

Three principal systems are used for the detection of X-rays:

a) Fluorescent screens.

They are based on layers of zinc sulphide doped with nickel. This compound has the property of X-ray fluorescence emission in the visible range, which can be detected by the human eye. Nevertheless, in order to reach an optimum detection level, it is necessary to have an X-ray incidence of high intensity. For this reason, this system of detection is not used when a cathode tube is employed as a source. However these fluorescent screens are utilised for the first calibrations of the equipments (5).

b) Counters:

These systems produce electric signal from an incident X-ray radiation. They are based on the X-ray ionisation property (6). Their applications come from their principal characteristics such as the efficiency, the losses and the resolution.

The counters can be classified in four groups: proportionals, Geiger, of excitation and of semiconductor. All of them have the attribute of detection from a minimum threshold and a detection efficiency which depends on the wavelength of the incident radiation, which must be known for rigorous studies of intensities with X-ray experiments. The advances in the counters detection technologies are, in general, directed: i) to obtain a signal when low radiation intensity arrives; ii) to work at room temperature; and iii) to keep control over the capability of temporal resolution.

c) Photographic films:

It must be pointed out that this kind of detectors will be discussed in detail in this section because they are the usual detection technique for the simple Laue-gram experimental set-up. Furthermore the knowledge of their characteristics is very important for understanding the results discussed in this book.

There are two kinds of film detectors: the standard photographic films, and the intensifying layers films.

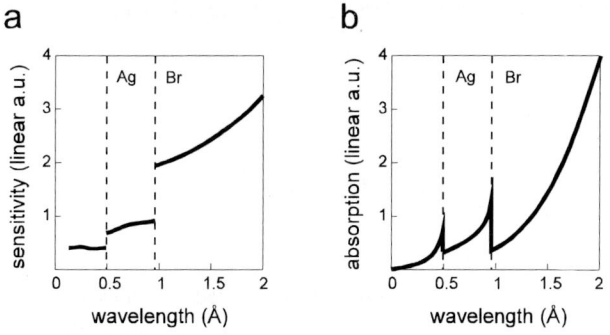

**Figure 1.4:** *Spectral sensitivity (a) and absorption (b) of an AgBr photographic film.*

*Standard photographic films* are based on the chemical emulsion properties of the photographic film. When a radiation falls on the film, a chemical reaction takes place which produces the blackness of the film. The level of darkness or film sensitivity produced depends on the wavelength of the incident radiation due to the different levels of chemical reaction with the emulsion for each wavelength. Therefore there is a spectral dependence of the film sensitivity. In general, this dependence is related with the absorption of the atoms forming the emulsion, which normally are Br and Ag. Figure 1.4 shows the spectral sensitivity and the absorption spectrum of the film. Although there is no direct correlation between the detection and the absorption (7), one can observe the relation of the sensitivity with the absorption edges of the Br (0.92Å) and the Ag (0.49Å). These kind of films are made with several emulsion layers on both sides of the film, in order to get a higher detection power. Furthermore the size of the emulsion film, in order to get a higher detection power. Furthermore the size of the emulsion grains generally is large enough to increase the total absorption and therefore the detection efficiency. However, the gain of efficiency due to the grain size reduces the spatial resolution. On the other hand, one must take into account the blackness saturation of these films whenever a study of intensities is to be carried out. In this case it is necessary to have in mind the exposure time during the experiment, and to know when the darkness reaches the saturation level. Furthermore, in order to know the blackness level of the film very precisely, one can use microdensitometers or microphotometers. Specifically, the Kodak AX films belong to these group, known as standard films. They have been mainly used for the discussion in this book.

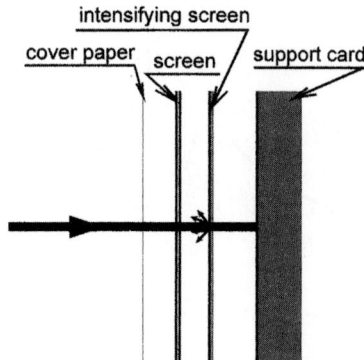

**Figure 1.5:** *Detection process of an intensifying film.*

The second kind of films are the *systems based on an intensifying layer*, which are mainly commercialised by Polaroid and are being employed in a large number of diffractometers due to their higher sensitivity and fast revealing process. Hence the cost and the exposure times of the experiments are reduced. Figure 1.5 shows a schematic diagram of these films, which are based on intensifying screen homogeneously spread over a support card. The chemical composition of this screen is either calcium tungstate for optimum detection of wavelengths of about 0.5Å, or zinc sulphur doped with silver when larger wavelengths are going to be used. Once the incident X-ray arrives at the chemical emulsion, a fluorescence is produced by the screen in the visible range which is detected by the film.

On these kind of films, the interval between the observable detection and the saturation level is much lower than on the standard films, and for this reason they are less useful for studies involving intensities analysis. Moreover, as the emission of the intensifying screen is in all directions, the perceptible effect is the enlargement of the diffraction spot, as more intensity is coming. This effect can be considered a criterion to evaluate the intensity levels. In spite of these limitations, these kind of films are normally used to carry out Laue-grams. Furthermore, in a large amount of Laue-gram applications, it is not necessary to work with the spot intensity in a precise way. However for a better simulation of Laue-grams, it is important to take into account the spectral sensitivity of the films in the X-ray range, which is currently missing for these intensifying films.

## 1.3 Crystal geometry

Atoms or ions arranged in a periodic and regular manner in three dimensions form a single crystalline material. A single crystal can be represented as a periodic geometrical assembly of spatial identical structural units called unit cell (8). The size, geometry and distribution of atoms within the unit cell depend on the specific material. Relevant nomenclature referred to crystal lattices, such as the Miller indices and the reciprocal lattice is going to be discussed.

### 1.3.1 Classification of crystallographic systems

Single crystals show the existence of certain symmetry elements, i.e. if certain operations are performed on them they can be brought to self-coincidence (9).

Attending to the unit cell parameters, there are, in all, seven crystal systems, taking into account the point symmetries, that is that each point have identical neighbours. These *7 crystal systems* are shown in figure 1.6, and any single crystal belongs to one of them. Other arrangements of points fulfil the requirements of a point symmetry: they are the so called *14 Bravais lattice* enumerated in table 1.1.

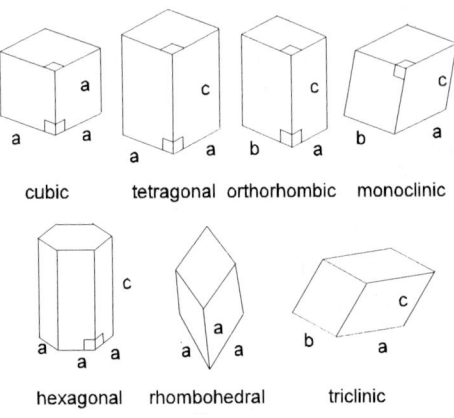

Figure 1.6: *The seven crystal systems.*

Table 1.1: *The fourteen Bravais lattices.*

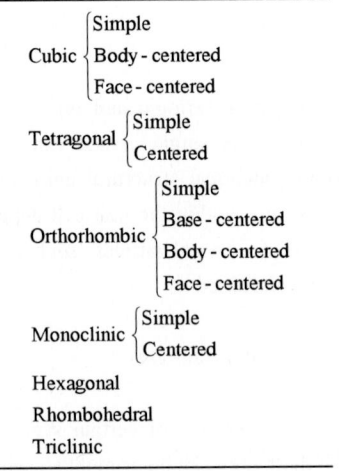

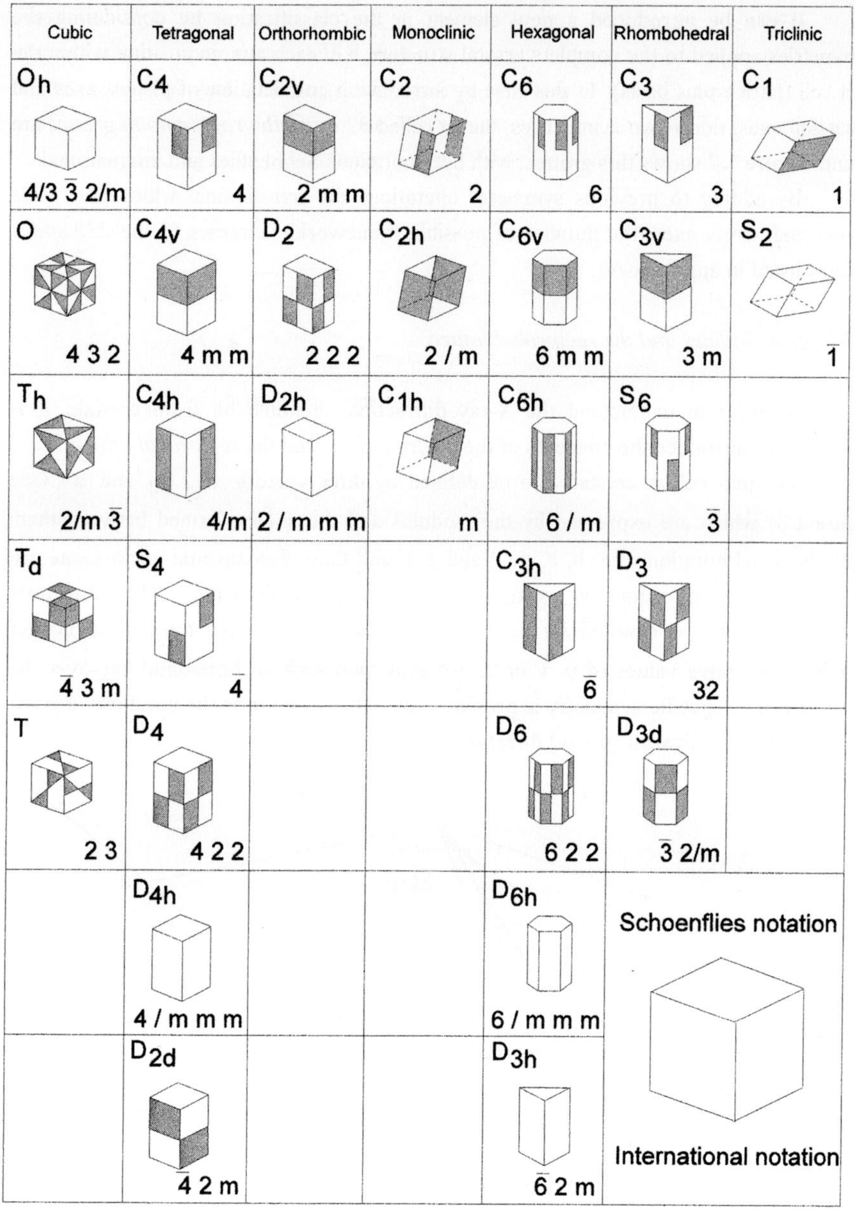

| Cubic | Tetragonal | Orthorhombic | Monoclinic | Hexagonal | Rhombohedral | Triclinic |
|---|---|---|---|---|---|---|
| $O_h$ | $C_4$ | $C_{2v}$ | $C_2$ | $C_6$ | $C_3$ | $C_1$ |
| 4/3 $\bar{3}$ 2/m        4 | 2 m m | 2 | 6 | 3 | 1 |
| O | $C_{4v}$ | $D_2$ | $C_{2h}$ | $C_{6v}$ | $C_{3v}$ | $S_2$ |
| 4 3 2 | 4 m m | 2 2 2 | 2 / m | 6 m m | 3 m | $\bar{1}$ |
| $T_h$ | $C_{4h}$ | $D_{2h}$ | $C_{1h}$ | $C_{6h}$ | $S_6$ | |
| 2/m $\bar{3}$ | 4/m | 2 / m m m | m | 6 / m | $\bar{3}$ | |
| $T_d$ | $S_4$ | | | $C_{3h}$ | $D_3$ | |
| $\bar{4}$ 3 m | $\bar{4}$ | | | $\bar{6}$ | 32 | |
| T | $D_4$ | | | $D_6$ | $D_{3d}$ | |
| 2 3 | 4 2 2 | | | 6 2 2 | $\bar{3}$ 2/m | |
| | $D_{4h}$ | | | $D_{6h}$ | Schoenflies notation | |
| | 4 / m m m | | | 6 / m m m | | |
| | $D_{2d}$ | | | $D_{3h}$ | International notation | |
| | $\bar{4}$ 2 m | | | $\bar{6}$ 2 m | | |

**Figure 1.7:** *The 32 crystallographic point groups.*

It can be introduced a new element in the classification by considering the symmetries applied to the complete crystal structure, i.e. each atomic position within the unit cell (lattice plus basis). In this case by introducing combination of planes, axes and inversion axes, thirty-two symmetries, the so called *32 crystallographic point groups* are found. Figure 1.7 shows this groups, with both notations Schoenflies and international.

By adding to previous symmetry operations the translations, which give glide planes and screw axes, the number of possible frameworks increases to the *230 space groups* listed in appendix A1.

### 1.3.2 Miller indices and the reciprocal lattice

In order to understand the X-ray diffraction phenomenon from crystals, it is important to introduce the concepts of the *Miller indices* and the *reciprocal lattice*.

The unit cell in crystals can be defined by three vectors, $\mathbf{a}_1$, $\mathbf{a}_2$ and $\mathbf{a}_3$ (see figure 1.8) which are expressed by the modulus and the angles formed between them with the usual notation of a, b, c, $\alpha$, $\beta$ and $\gamma$. Using these fundamental vectors one can designate any direction in space by a linear combination. The direction, $u\mathbf{a}_1 + v\mathbf{a}_2 + w\mathbf{a}_3$, is denoted by [u v w] where u, v and w are normalized integer numbers. Negative values of u, v or w are expressed with an horizontal bar over the number. If any specific symmetry is present in the lattice, one uses the notation <u v w> to lump all the equivalent [u v w] directions.

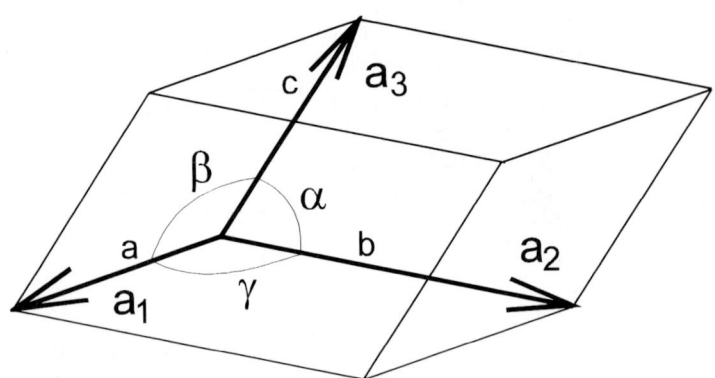

**Figure 1.8:** *Scheme of an unit cell defined by the vectors of the direct lattice $a_1$, $a_2$ and $a_3$ (with modulus a, b and c respectively) showing the angle nomenclature.*

The Miller indices (h k l) are used to denote the lattice planes. Considering a known unit cell with direct lattice vectors $\mathbf{a}_1$, $\mathbf{a}_2$, $\mathbf{a}_3$, the plane (h k l) is the one which goes through the points $\mathbf{a}_1/h$, $\mathbf{a}_2/k$ and $\mathbf{a}_3/l$. It must be pointed out that a zero value for an index means that the plane is parallel to the vector (see figure 1.9). The plane ( nh nk nl ), with an integer number n, has the same orientation of that of the plane (h k l), although the spacing is n times lower in the former case. This consideration will be useful for the treatment carried out during the discussion of the intensity of the diffracted beams (section 2.3).

When an hexagonal system is considered, a new index (t) is included, for a better visualisation of the symmetry. The value of t is given by $t = -l(h + k)$, and the complete notation will be written as (h k t l).

When for a given symmetry there are several planes with the same characteristics, for example the planes (100), (010), ($\overline{1}$00) and (0$\overline{1}$0) in a monoatomic tetragonal structure, one can refer to all of them using the notation {100}.

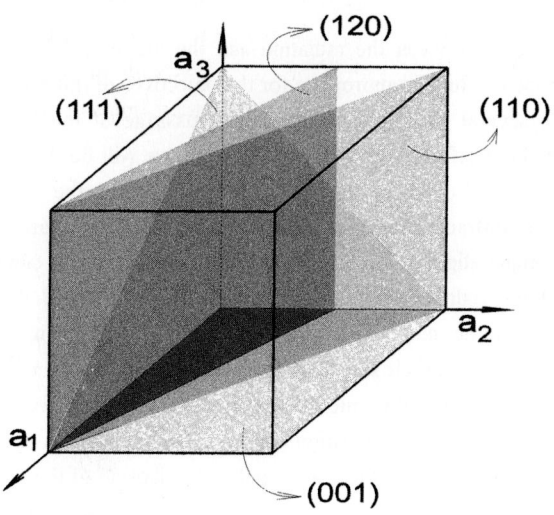

**Figure 1.9:** *Examples of lattice planes labelled with their Miller indices.*

The most complete information about the unit cell is given by the so called reciprocal lattice. The perpendicular direction to each of the planes (h k l), and the interplanar spacing are very useful information in crystallography. The  vectors of the reciprocal lattice, $b_1$, $b_2$ and $b_3$, are defined as:

$$b_1 = a_2 \wedge a_3 / \Omega \qquad b_2 = a_3 \wedge a_1 / \Omega \qquad b_3 = a_1 \wedge a_2 / \Omega$$

where $\Omega$ is the volume of the unit cell.

By geometrical considerations it is found that:

i) the direction of the vector $hb_1 + kb_2 + lb_3$ is perpendicular to the (h k l) plane, and

ii) the modulus $|hb_1 + kb_2 + lb_3|$ is the distance between the (h k l) planes of the lattice.


## 1.4 Scattering by an atom

The interaction between the radiation and the matter will be discussed from a classical point of view, which is appropiate for the objectives of this book.

There are several effects produced by the passage of the X-rays through the matter. Amongst these effects, only the scattered X-rays will be discussed here, which can be either coherent or incoherent. Out of these, only the coherent scattering contributes to the diffracted beam since all the scattered radiation has the same wavelength. A simple dipolar model is used to understand the coherent scattering. Keeping in mind the tight bonding of the electron, the  spectral distribution of the scattered radiation is rather narrow and close to the incident radiation. For a given ion or atom, the contribution of all their surrounding electrons must be also taken into account.

The scattering produced by the atom or ion is isotropic, if we assume that i) the electron which is affected by the radiation can be in any initial state and  ii) the incident wave is unpolarized. In practice,  there is a  spatial distribution of the emission as will be discussed in section 2.3.6.

Interference phenomenon is observed when two spatially and temporally coherent fields interact. This condition is satisfied when waves of the same wavelengths are involved. Figure 1.10 shows the situation of two electrons irradiated by a collimated beam. Considering this geometry, when two incident sinusoidal waves of the same wavelength are reaching two electrons A and B, a coherent scattering can be produced in any particular direction. Let us consider the direction AY for the scattered waves coming from the electrons A and B. The angle $2\theta$ will be defined as the one formed between the incident and emitted beam. The wave, which reaches the electron B, produces a coherent radiation, and its path is lower than the path of the one emitted by the electron A. The resulting wave due to interference is given by the addition of the two individual fields. An example of two out of phase sinusoidal waves with amplitudes $A_1$ and $A_2$ is shown in figure 1.11. A totally destructive result is achieved for a phase difference of $\pi$, while a zero out of phase means a totally constructive condition. The phase difference can be derived from the path difference between the two waves. The intensity of the resultant wave after the interference process depends on the ratio of $\sin\theta$ to $\lambda$.

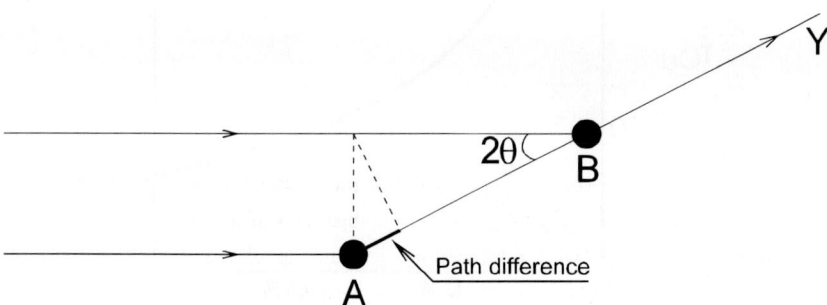

**Figure 1.10:** *Interference scheme of two electrons which are scattering coherently.*

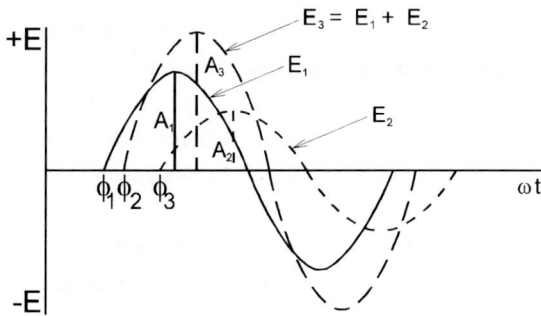

**Figure 1.11:** *Composition of two out of phase sinusoidal electric fields.*

Till now, an arbitrary fixed geometry for the electrons have been considered. Nevertheless, following the approximations of Hartree-Fock (10) or Dirac-Slater (11), the statistical positions of the electrons can be estimated. This is calculated and normalised for each elements by the form factor f, which is related with its electronic distribution. As an example, figure 1.12 shows the behaviour of the copper (Cu) form factor on $\sin\theta/\lambda$.

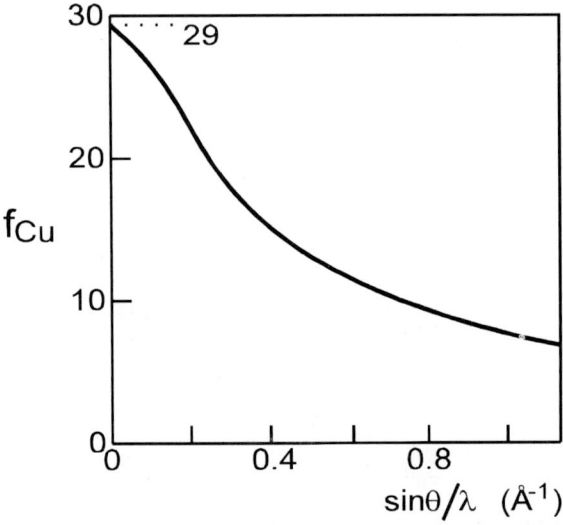

**Figure 1.12:** *Dependence on $\sin\theta/\lambda$ of the copper atomic scattering form factor ($f_{Cu}$).*

## 1.5 Diffraction by lattices and the Bragg law

In this section, the diffraction by lattices is discussed by considering the scattering by only an atom or an ion for each element of the unit cell i.e. one of the lattice of the seven crystal systems. There can be many atoms per unit cell and their positions are well known for any resolved structure. By means of the structure factor (F), the total contribution of the atoms of an unit cell to the diffraction intensity is accounted, as will be discussed latter in section 2.3.3.

Let us consider a group of atoms situated in an periodical structure as shown in figure 1.13, and at the same time an incident collimated X-ray radiation coming on this structure. If two scattered directions are taken into account from the atoms A and B, both with the intensity depending on the form factor (f) introduced in the previous section, there is a path difference between the two interfering waves. This path difference depends on the distance between the atoms and on the function sinθ, giving rise to an out of phase condition depending on the wavelength value.

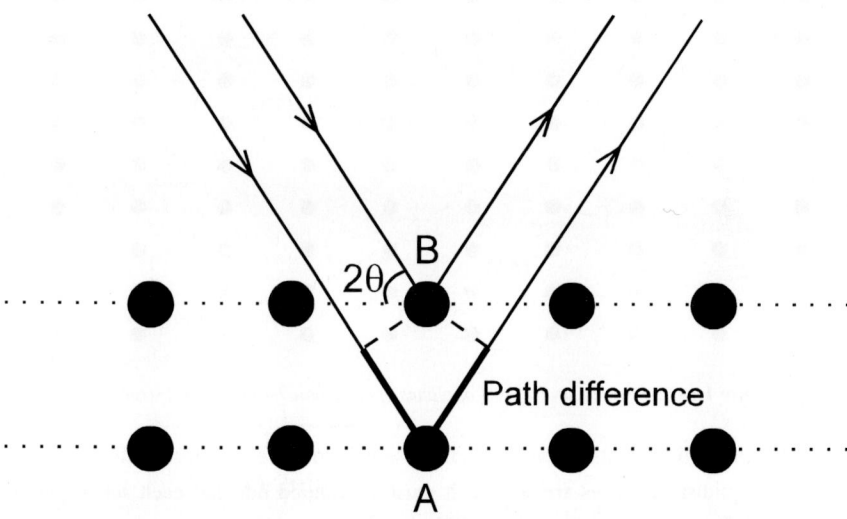

**Figure 1.13:** *Scheme of the path difference between two beams scattered by two adjacent atoms exposed to a collimated radiation.*

Considering that there are a large number of periodic units which are being irradiated, the diffracted beams are composed of a large number of scattered rays interfering with totally constructive condition in time and position. The occurrence of time coherence is assured due to the fact that the interference processes take place within a time of $10^{-10}$ seconds in a 3 cm sample-detector distance experiment.

The physical phenomenon of diffraction by crystalline lattices can be easily understood from the Bragg law. It arises from the fact that for a given incident direction, and considering a known periodic distribution of the scattering elements and a fixed geometry of the system, the wavelengths that satisfy the totally constructive interference condition are perfectly known.

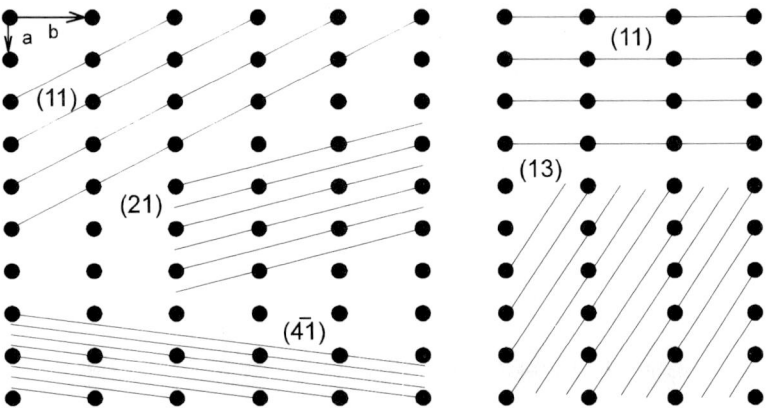

**Figure 1.14:** *Scheme of several sets of planes of a periodic bidimensional structure.*

Figure 1.14 presents a periodic bidimensional structure on which different groups of parallel equidistant planes are shown. It must be pointed out that each set of parallel planes has different density of units and different interplanar distance.

In fact, there is a group of planes normal to any given direction. By considering all the possible groups of equidistant planes, one can get the diffraction condition for each of them as will be discussed in the next paragraphs.

Let us consider the situation shown in figure 1.15, in which two waves (I) fall on two consecutive planes with an angle $\theta$. Both waves reflect (R) with an angle $\theta$ too. The path difference between the two waves is $2d \sin\theta$, the interference being totally constructive when it is equal to an integer number times the wavelength of the radiation. Therefore the resulting expression is $n\lambda = 2d \sin\theta$, where n is an integer number.

Furthermore, considering the same incident radiation and other emission direction, there will be another group of parallel planes with a different d' as distance between them and θ' as the incident angle, for which the diffraction condition will be nλ = 2d' sinθ .

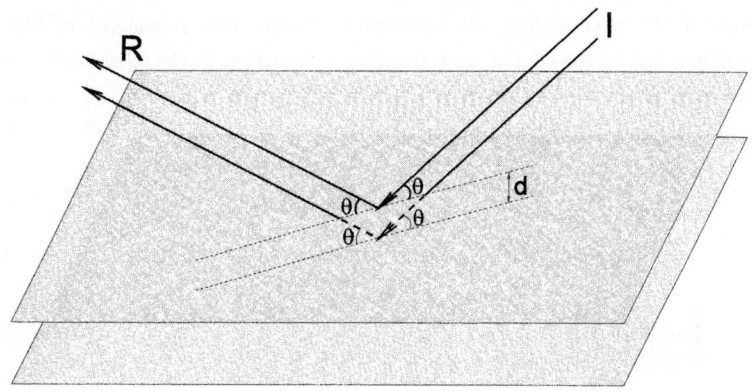

**Figure 1.15:** *Reflection by two parallel planes of two falling parallel beams.*

From previous discussions, it is practical and convenient to consider any diffraction direction as the reflected one by a determined set of planes. By analogy the order diffraction values n = 2, 3, 4, ... can be represented as the effective diffraction by 2, 3, 4 ... intercalated fictitious planes within the set.

Previous condition for the diffraction phenomenon is the well known Bragg law (12) which presents the same information that of the von Laue vector equations which are derived from the reciprocal lattice (13). The Bragg law is very useful for calculating the wavelengths and directions for which the diffraction condition is fulfilled by any given interplanar distance.

It is worth noting that, for a condition where, sinθ ≤ 1, and hence λ/2d ≤ 1, the diffraction condition implies that λ ≤ 2d. For a wavelength λ << 2d, the incident angle θ which satisfies the diffraction condition must be extremely low. For this reason, to get the diffraction condition with lattice parameters in the order of Å, it is necessary to use a radiation of wavelength λ ~ Å.

Lowest values of X-ray wavelengths, $\lambda_{min}$, from conventional sources are around 0.5Å. Therefore from the previous limit of $\lambda_{min} \leq 2d$, there are a selected discrete number of set of planes which are able to fulfil the Bragg condition. These are the ones with interplanar distance with a value of $d \geq \lambda_{min} / 2$.

If the Miller indices (h k l) are used to define the planes of a given lattice (as shown in figure 1.14), it can be demonstrated that the higher the index numbers are, the lower the interplanar distance d. Without taking into account other factors which will be considered in the next section, the resultant directions from reflections of low index number planes can be considered, as a rough approximation, as the most intense ones. In the same way, the number of directions able to diffract are higher for larger parameters of the unit cell (larger interplanar distance).

# CHAPTER 2

## BACK-REFLECTION LAUE TECHNIQUE

In this chapter, we first present the experimental set-up for the X-ray back-reflection Laue technique. The main applications of this method are mentioned. Finally the main factors which determine the intensity in a given diffraction pattern are deduced and described in detail.

## 2.1 Geometry of the back-reflection Laue technique

The geometry of the back-reflection Laue technique is shown in figure 2.1. Basically, the experimental set-up has the following components: an X-ray tube where radiation goes through a pinhole collimator (C), a photographic plate holder (H) situated at a distance s from the sample (S) which is mounted on a goniometer holder (B).

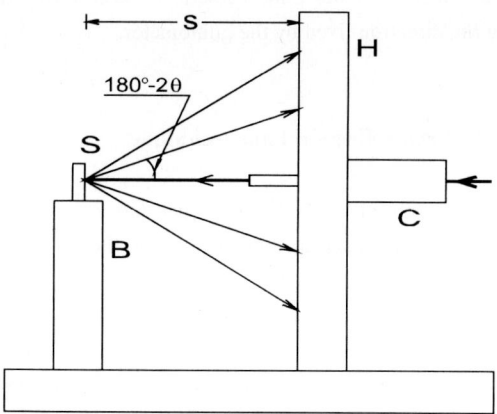

**Figure 2.1:** *Geometry of the back-reflection Laue technique.*

The X-ray tube is situated in such a way that its window is aligned with the centre of the plate-holder, which in turn is aligned with the eucentric point of the goniometer. A good alignment of this system is important for an accurate result. The most usual procedure for alignment is done using a fluorescent screen, looking for a zero eccentricity of the ellipse which results from the radiation falling on the plate. When a more accurate orientation is required (accordingly with the precision of the goniometer) a laser must be used for the alignment (14).

In order to achieve the symmetry in the diffraction Laue pattern, the sample surface opposite to the X-ray incident radiation must be totally perpendicular. Otherwise, even with a right orientation, absorption effects can modify the symmetry relations of the spot intensities. This condition has been assumed in the calculations developed in section 2.3.7. For this reason, it is advisable to lap the sample prior to each back-reflection Laue experiment.

Another important requirement in an experimental orientation work, is the sample-detector distance (s) in order to calculate the diffraction direction projections on the detector plate, as will be discussed latter in section 3.3.3. This fact implies that the distance between the detector and the so called volume diffraction centre of the sample must be carefully estimated. This centre depends on the material absorption; for a larger absorption of the sample, nearer is the volume diffraction centre to the sample surface.

To carry out any orientation experiments from a previous Laue pattern, it is mandatory to fix the position of the diffraction centre at the eucentric point of the goniometer. Hence, it is advisable to work with a lapping and cutting machine which is able to mount the goniometers of the Laue camera, in such a way that the cutting or lapping is done along the direction fixed by the goniometer.

## 2.2 Applications of the back-reflection Laue technique

Amongst the diffraction methods, the Laue patterns are the easiest to obtain and requires the simplest experimental set-up. For this reason, they are used as a first test to check the monocrystallinity of the materials.

This technique uses a white radiation as produced by the cathodic source.

As a consequence of the continuum in the wavelength which falls on the sample, a lot of diffraction conditions are produced. The symmetries of the crystalline structures can be recognized from the projections registered on the Laue-gram (15). Therefore, this technique is used extensively to get the orientation of single crystals.

Laue-grams also record diffuse scattering originating from the thermal vibrations around the atomic positions in the crystal (see section 2.3.4). The back-reflection Laue patterns register this information in a more reliable way than the transmission one (16,17). However, the large diffusive scattering can be an obstacle, when the main objective of the experiment is just to get the information from diffraction.

The polychromaticity and hence, the large number of diffraction conditions are the inherent properties of the Laue technique. This polychromaticity is a disadvantage when further analysis of the pattern is required. Specific information can be obtained in a simpler manner by employing monochromatic methods.

However, taking into account that in the Laue technique there is no movement of the sample, and the exposure times are shorter than in other techniques: methods of data analysis have been developed for Laue patterns which include the intensity calculations (18). Using this procedures, the structures of small size samples, on which eucentric rotations are not possible, can be studied. Also complex organic structures like proteins, which require extremely short exposure times can be studied by recording the Laue patterns without damaging the sample, by employing synchrotron sources.

The back-reflection technique has the advantage that sample preparation is not required, unlike in the transmission geometry, where a critical thickness of the sample cannot be exceeded due to absorption considerations. For this reason, bulk samples can be easily oriented with the back-reflection Laue technique.

## 2.3 Factors which determine the diffraction intensities in the back-reflection Laue patterns

In this section, the main factors and circumstances which influence the intensity in a back-reflection Laue pattern will be discussed. Latter they will be included during the simulation of Laue patterns which considers the spots intensities (section 3.2).

### 2.3.1 Incident radiation

To carry out a relative estimation of the intensities in a Laue pattern, it is necessary to know the spectral distribution of the incident radiation which falls on the sample.

For an orientation work, as previously discussed in section 2.2, the advantages of a Laue pattern arises from a homogeneous incident spectral distribution. As discussed in section 1.1, if we are working with a cathodic source, a voltage higher than the threshold one, above which the characteristic lines of the material used as anode appear, should be avoided.

If we exceed this voltage, the use of this emission will produce very few highly intense spots in the pattern, which obey the Bragg condition with the wavelength of the characteristic lines. These intense spots will dictate the exposure time and the number of diffraction spots detectable on the patterns, which leads to a reduction in the information. Some indexing and simulation models use this aspect of characteristics lines (19), due to the simplicity of the analysis of monochromatic data that can be performed in these cases. However, the low number of spots recorded leads to a difficulty in    final identification of the experimental Laue-grams from the solutions proposed by the identification algorithms.

In the model proposed in this book, the intensities of the patterns are going to be estimated in a relative mode. The current intensity applied to the cathodic tube enhances the emission by an equal proportion for each wavelength. The current intensity will not be  required for the simulation work. However it must be pointed out that there is an inverse ratio between the current intensity and the exposure time.

When a filter or an encapsulant is used, it is necessary to know its thickness and the transmittance spectral dependence, in order to estimate the actual intensity which falls on the sample. However, other experimental accessories such as pinholes can be introduced without special care for our simulation comparison, due to the relative nature of our model.

### 2.3.2 Film detector

Once the beam has been diffracted and reaches the detector film, one must take into account a spectral dependence on the film sensitivity and therefore, not all the wavelengths are detected with the same efficiency, as discussed in section 1.2.

There is another important point which must be considered: the two threshold levels of the detector film. The minimum threshold and the saturation level. Below the minimum threshold one cannot identify a spot in the film. Above the saturation level, higher intensities cannot be registered, because of the maximum blackening has been reached in the film.

Following these considerations, the intensity values are influenced by several specific details, such as the exposure time, the operating time of the tube used as source, the sample size, the sample absorption, the sample diffraction power, the cross section of the radiation beam, etc. From the relative nature of the model proposed in this book, these levels must be fixed in the simulated Laue-gram by comparison with the experimental one.

### 2.3.3 Structure factor (F)

The crystal structure, which is determined by the lattice parameters and the atomic positions within the unit cell, is an essential element for the diffraction intensities. Actually, the principal application of the X-ray diffraction comes from the fact that one can determine crystal structures, mainly by powder techniques.

Each atom scatters according to its electronic distribution following a dependence on $\sin\theta/\lambda$, as previously discussed in section 1.4. The form factor (f) gives us this dependence, and is well calculated for all elements and some ions (2.b).

Taking into account a general instance in which there is more than one atom per unit cell, the interference between the waves scattered by each atom is considered as the diffraction unit. The way to determine it, is by means of the so called structure factor F, which is the addition of the waves scattered by each of the atoms of the unit cell, considering that there is coherence in time and space. Considering that we are interested in relative intensities, the proportional factors are going to be neglected for the following discussion.

Let us consider the unit cell (with parameters a, b and c) shown in figure 2.2, with two atoms at positions 1 and 2, their relative coordinates being $x_1 = 0/a$, $y_1 = 0/b$, $z_1 = 0/c$, and $x_2 = u/a$, $y_2 = v/b$, $z_2 = w/c$. Let us take into account a diffraction direction which corresponds to the reflection of a particular (h k l) plane in the Bragg condition.

With this consideration, the path difference ($\delta$) between the waves of wavelength $\lambda$ scattered by the atoms 1 and 2 is given by

$$\delta = \lambda\left(\frac{u}{a/h} + \frac{v}{b/k} + \frac{w}{c/l}\right).$$

Hence the phase difference is $\phi = 2\pi\delta/\lambda = 2\pi(hx_2 + ky_2 + lz_2)$.

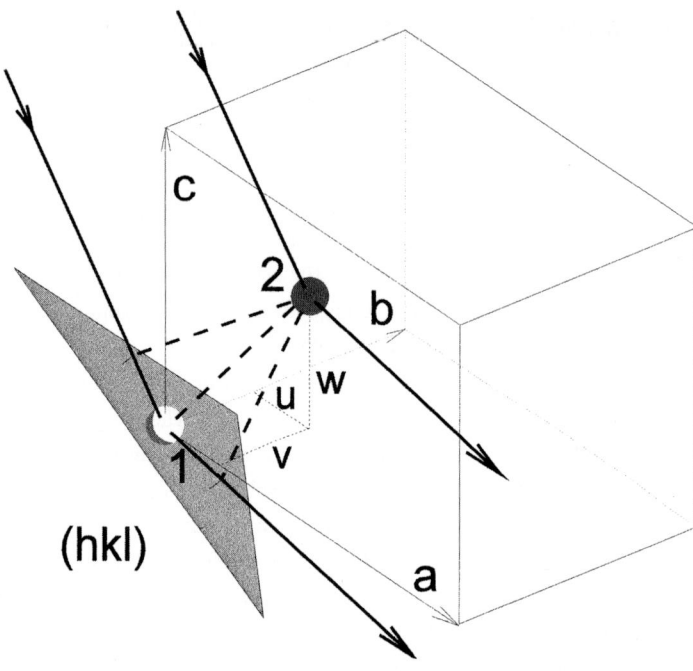

**Figure 2.2:** *Three dimensional scheme of two atoms of an unit cell, 1 in (0,0,0) and 2 in (u,v,w), showing the path difference between two parallel diffracted rays by the (h k l) plane.*

By taking into account the equation [1] in section 1.1, and setting a zero phase at the position of atom 1, we can write the electric fields of the two waves which obey the following proportionalities:

$$\mathbf{E}_1 \propto A_1\, e^{-i\, 2\pi vt} \qquad\qquad \mathbf{E}_2 \propto A_2\, e^{-i\,(2\pi vt - \phi)}.$$

Removing the time proportionality factor,

$$\mathbf{E}_1 \propto A_1 \qquad\qquad \mathbf{E}_2 \propto A_2\, e^{i\phi}.$$

Furthermore, known that each amplitude is proportional to the form factor f.

$$\mathbf{E}_1 \propto f_1 \qquad\qquad \mathbf{E}_2 \propto f_2\, e^{i\phi}.$$

Replacing the value of the phase difference $\phi$,

$$E_1 \propto f_1 \qquad\qquad E_2 \propto f_2\, e^{2\pi i(hx_2+ky_2+lz_2)}.$$

In a general case, where instead of two, there are n atoms per unit cell, the proportional expression for the resultant amplitude of the electric field (F), will be given by,

$$F = f_1 e^{2\pi i(hx_1+ky_1+lz_1)} + f_2 e^{2\pi i(hx_2+ky_2+lz_2)} + ... + f_n e^{2\pi i(hx_n+ky_n+lz_n)}$$

which is known as the structure factor (F).

The intensity of the resulting field (I), is proportional to $|F|^2$. In the trigonometric form, the final expression becomes,

$$|F|^2 = \left[ f_1 \cos 2\pi(hx_1 + ky_1 + lz_1) + \; ... \; + f_n \cos 2\pi(hx_n + ky_n + lz_n) \right]^2 +$$
$$\left[ f_1 \sin 2\pi(hx_1 + ky_1 + lz_1) + \; ... \; + f_n \sin 2\pi(hx_n + ky_n + lz_n) \right]^2.$$

### 2.3.4 Temperature factor

As a first approximation, it was considered that the atoms within the unit cell occupy static positions. However, thermal oscillations with an amplitude (u) occur around them. Therefore, since the distances between atoms, and hence between planes, are related with the mean oscillation amplitude, the structure factor expression must be modified accordingly. These oscillations induce changes in the path difference between the diffracted waves. This is enhanced when the oscillation amplitude (u) is large compared to the interplanar distance (d) and therefore it will be more evident for smaller d, i.e. larger Miller indices. Since $d = \lambda / 2 \sin\theta$, the effect of the temperature-diffuse scattering is larger for higher incident angles, which gives the central ring present in the experimental back-reflection Laue-grams.

The value of u, is a function of the temperature dependence of the elastic constants of the material. The way to include this effect on the diffraction spot intensities estimation is by correcting the form factor (f) of each atom, by multiplying the temperature factor, $e^{-M}$. The value of M (20) contains the dependence on $(u/d)^2$, through the mean value of the square displacement, $\overline{u^2}/d^2$,

$$M = 2\pi\left(\frac{\overline{u^2}}{d^2}\right) = 8\pi^2\overline{u^2}\left(\frac{\sin\theta}{\lambda}\right)^2.$$

The rigorous calculation of $\overline{u^2}$ is difficult, because it is required a previous knowledge of the material properties like its elastic constant. The Debye theory of specific heats will be used in the model employed in this book. This procedure takes into account a linear approximation in the dispersion branches of the crystal and a spherical approximation for the reciprocal space of the lattice, with a fixed radius according to the Debye temperature of the material $\Theta_D$. Over this fixed temperature, it is assumed that all the vibration modes of the phonons are activated. With this approximation, the following expression for M is derived (21),

$$M = \frac{6h^2T}{mk_B\Theta_D^2}\left[\varphi\left(\frac{\Theta_D}{T}\right) + \frac{\Theta_D}{4T}\right]\left(\frac{\sin\theta}{\lambda}\right)^2$$

where h is the Planck's constant, m is the mass of the vibrating atom, $k_B$ is the Boltzmann constant, T the operating temperature and $\varphi$ the Debye function which has the following expression,

$$\varphi(x) = \frac{1}{x}\int_0^x \frac{\xi}{e^\xi - 1}d\xi$$

Although this Debye theory was enunciated for a cubic crystal, in our model the Debye expression will be applied to any crystal structure. This extension is a rough approximation. Nevertheless, it can be assumed for our work. It is worth mentioning that the Debye temperatures of non-cubic crystals can be found in the literature.

### 2.3.5 Anomalous dispersion

In the previous sections the dependence of the form factor on the ratio $\sin\theta/\lambda$ was introduced, which is a normalised value related with the electronic distribution of each element. However, when the atomic number of the element is high enough, and the energies of its K lines have a value close to the incident X-ray, a resonance effect is produced with the electrons associated to the K transitions at the wavelength $\lambda_K$. This phenomenon is known as the anomalous dispersion and consequently will modify the value of the form factor.

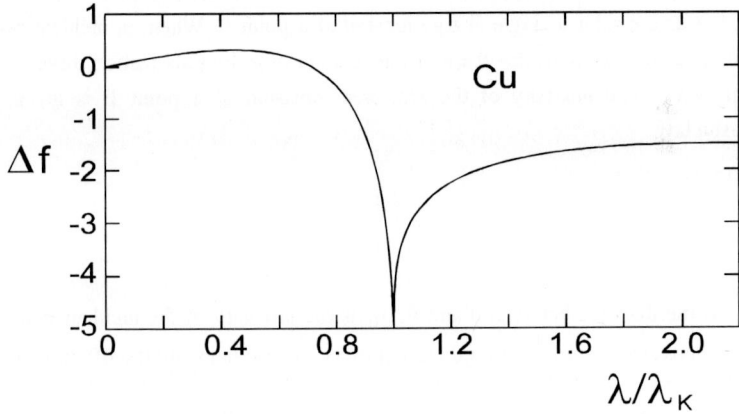

**Figure 2.3:** *Copper scattering form factor correction, $\Delta f$, in the vicinity of the K absorption wavelength ($\lambda_K$).*

Previous calculations for the form factor (section 1.4) were made assuming values of the incident wavelengths far away from the one which produces the K absorption of the material. Therefore a correction value $\Delta f$ must be added. Figure 2.3 shows the calculated $\Delta f$ value for the Cu atom in the vicinity of its $\lambda_{K,Cu}$. Due to the fact that the value of the $\Delta f$ corrections are slightly dependent on the atomic number Z of the scattering element, the calculated corrections of the Cu for any atom is generally used (considering the error magnitude assumed in most of the X-ray works). The Cu atom is used as master correction because of the mean value of its electronic population.

It must be pointed out that the anomalous dispersion is used as a tool in some X-ray spectroscopies, because it is possible to discern the diffraction contributions of similar elements by working with an incident wavelength close to the K transition wavelength $\lambda_K$ of one of them.

### 2.3.6 Polarization factor

At the moment, we have not taken into account any consideration about the polarization of the X-rays which falls on the material. Therefore, the re-emission probability for a given direction was discussed neglecting this fact. Nevertheless, considering a classical oscillating model, it can be demonstrated that there exists an angular dependence in this probability (22), which depends with the angle between the incident and the re-emitted directions.

Let us consider a single charge located at a point 0. When an incident polarized radiation falls at this point, the charge oscillates obeying the polarization direction of the incident wave. The intensity of the scattered emission at a point P is given by the Thompson law,

$$I_P \propto \frac{I_0}{r^2} \sin^2 \tau$$

where r is the distance between 0 and P, $I_0$ is the intensity of the incident wave, and $\tau$ the angle between the polarization direction of its electric field and the 0P direction.

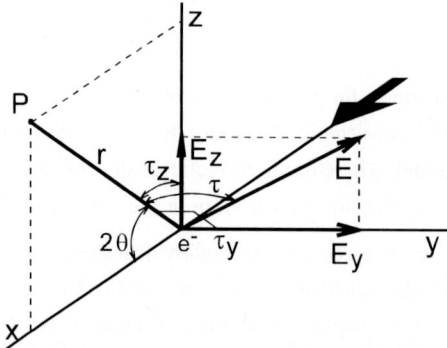

**Figure 2.4:** *Scheme employed for the calculation of the radiation intensity at P produced by the scattering of a single electron.*

In the general case shown in figure 2.4, an unpolarized wave advances on the x axis and falls on an electron located at the origin of the reference system. In order to know the intensity which falls at a point P of a given scattering direction, one has to take into account that the unpolarized wave fields fulfil the relation,

$$\mathbf{E}_y^2 = \mathbf{E}_z^2 = 1/2\ \mathbf{E}^2$$

hence,

$$I_{0y} = I_{0z} = 1/2\ I_0$$

The y component of the electric field accelerates the charge in the 0Y direction, while the z component does it in the 0Z direction. Considering that (see figure 2.4) $\tau_z + 2\theta = \pi/2$, $\tau_y = \pi/2$ and that the proportionality factors are neglected, the total intensity ($I_P$) will be given by the composition of

$$I_{Py} \propto \frac{I_{0y}}{r^2} \qquad \text{and} \qquad I_{Pz} \propto \frac{I_{0z}}{r^2}\cos^2 2\theta,$$

therefore,

$$I_P \propto \frac{I_0}{r^2}\left(\frac{1+\cos^2 2\theta}{2}\right).$$

The factor $\frac{1}{2}(1+\cos^2 2\theta)$ is named polarization factor, and must be included in all the calculation for diffraction intensities, when an unpolarized incident radiation is used.

### 2.3.7 Absorption factor

Taken into account all the interaction processes between the radiation and the material, one must consider the loss of intensity when a beam goes through a sample. This loss is lumped in a macroscopic way by the linear absorption coefficient ($\mu$) of the material, following an exponential relation related with the thickness of the sample which is crossed by the radiation, which is given by the expression $I_x = I_0 \, e^{-\mu x}$.

Figure 2.5 shows a sketch of the back-reflection geometry that is going to be used to obtain the correcting absorption factor (AF) for this technique. In this figure the incident radiation is perpendicular to the external face of the material, a condition which will be assumed in the proposed simulation model.

Let us consider an incident collimated beam with an unity circular cross section of radius a ($\pi a^2 = 1$) and with an intensity per unit area $I_0$. For a given (h k l) plane located at a depth AB, the arriving intensity is $I_0 \, e^{-\mu(AB)}$.

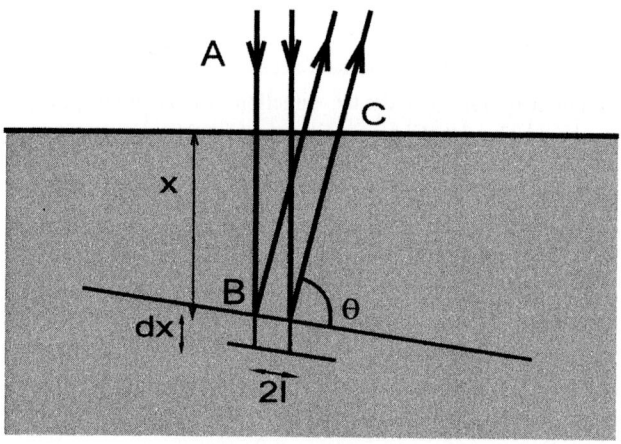

**Figure 2.5:** *Path of an incident and diffracted beam along a sample in a back-reflection geometry.*

Considering the fraction of the incident energy which is diffracted (p), the total intensity which is diffracted by the plane (h k l) is: $p \, I_0 \, e^{-\mu(AB)}$. Therefore the intensity per unit area back-reflected is $p \, I_0 \, e^{-\mu(AB)} \, \pi ab$, where 'b' is the semi-axis of the ellipse of the cross section of the diffracted beam generated by the intersection between the incident plane and the reflecting plane. It must be pointed out that the minor semi-axis 'a' is equal to the initial radius of the incident circular cross section. Finally, the intensity of the diffracted beam which emerges from the external face is, $p \, I_0 \, e^{-\mu(AB)} \, \pi ab \, e^{-\mu(BC)}$.

Taken into account the contribution of a differential volume of planes (h k l), it can be shown $dI = p\,I_0\,e^{-\mu(AB)}\,\pi ab\,e^{-\mu(BC)}\,dx$. Due to the relative nature of the proposed model, we can neglect the constant factors in the previous expression. Hence $dI \propto e^{-\mu(AB+CD)}\,b\,dx$. Using geometrical considerations from the figure, it is deduced that AB=x, $BC = x / \cos(\pi - 2\theta)$ and $b = 1/\cos(\pi/2 - \theta)$, $\theta$ being the reflection angle.

Integrating the contribution of the whole volume of diffraction, the total intensity of the diffracted beam, $I_D$, follows the expression,

$$I_D \propto \int_{x=0}^{x=\infty} e^{-\mu x\left(1+\frac{1}{\cos(\pi-2\theta)}\right)}\frac{1}{\cos(\pi/2-\theta)}\,dx\,.$$

In this integral, extending the integration volume to $\infty$ is an approximation with fairly good accuracy for back-reflection Laue experiment because the order of magnitude of $\mu$ values and the behaviour of the exponential function.

By integration of the preceding expression, we found the absorption factor (AF) which will be used in the next simulation model (section 3.2.1) where the spectral dependence of the linear absorption factor is written as

$$I_D \propto \frac{1}{\mu(\lambda)\left(\cos(\pi/2-\theta)+\dfrac{\cos(\pi/2-\theta)}{\cos2(\pi/2-\theta)}\right)} = AF\,.$$

### 2.3.8 Lorentz factor

A perfect collimation of the incident beam was assumed in all our previous discussions. Nevertheless, there exists an angular de-collimation in the incident X-ray beam. These small variations give rise to an angular distribution in the intensity of the diffracted beam, as shown in figure 2.6.

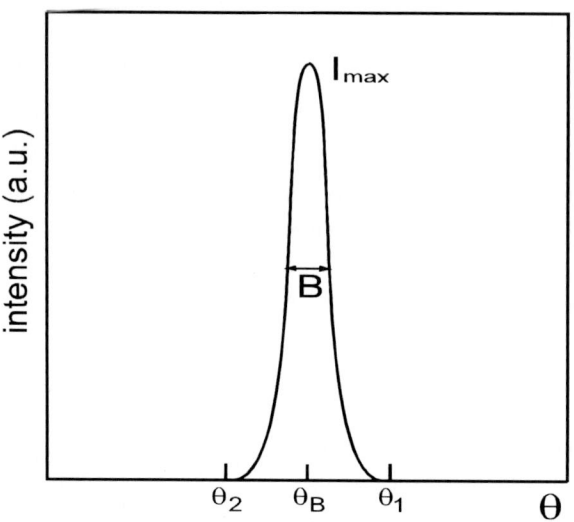

**Figure 2.6:** *Angular distribution of a diffracted beam around the Bragg angle $\theta_B$.*

The Lorentz factor (LF) integrates the total intensity value considering all the geometrical aspects which produce the angular broadening. This integration is approximated by multiplying the full width at half maximum (B) by the maximum intensity value $I_{max}$ (see figure 2.6).

In order to estimate the dependence of B, for a given number of planes, one must consider a range around the incident angle from $\theta_1 = \theta_B + \Delta\theta$ to $\theta_2 = \theta_B - \Delta\theta$, where $\theta_B$ is the angle that fulfil the Bragg condition for a given wavelength ($\lambda$), as it is shown in figure 2.7.

When the incident angle $\theta_1$ is considered, there is a small phase difference between the beams reflected by the planes 0 and 1. However, there is a deeper plane, m+1, for which the phase difference between 0 and m+1 is equal to one wavelength. It means that the intermediate planes are cancelling its diffractions between themselves. In the same way it happens for the angle $\theta_2$, although the cancelling condition will occurs at the plane $m-1$. Let us consider from the first to the last reflecting plane at a depth $t = dm \approx d(m-1) \approx d(m+1)$, where d is the interplanar spacing. The limit values for $\theta_1$ and $\theta_2$ satisfy the equations:

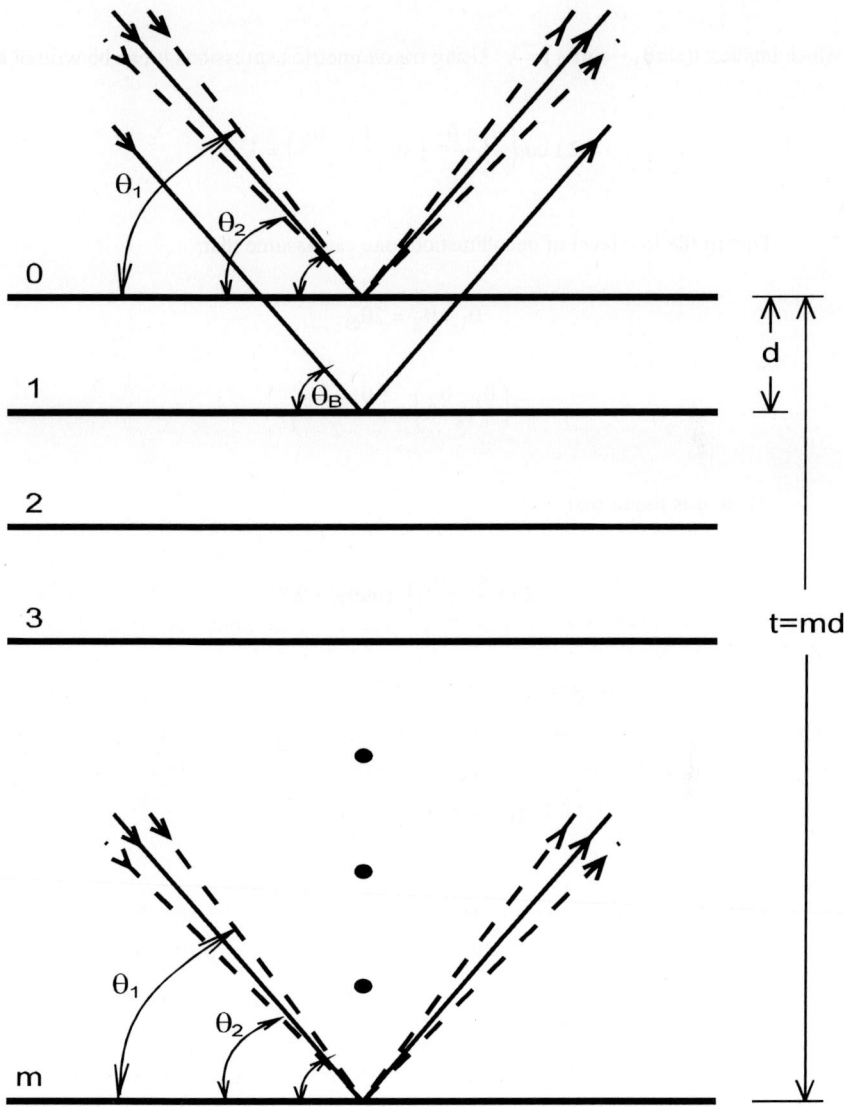

**Figure 2.7:** *Effect of de-collimation on angular distribution of diffraction by bulk crystals.*

$$2\,d\,\sin\theta_1 = \lambda \left.\right\} \quad 2\,t\,\sin\theta_1 = (m+1)\lambda \left.\right\}$$
$$2\,d\,\sin\theta_2 = \lambda \left.\right\} \quad 2\,t\,\sin\theta_2 = (m-1)\lambda \left.\right\}$$

which implies $t(\sin\theta_1 - \sin\theta_2) = \lambda$. Using trigonometric expressions it can be written as

$$2\,t\,\cos\left(\frac{\theta_1+\theta_2}{2}\right)\sin\left(\frac{\theta_1-\theta_2}{2}\right) = \lambda.$$

Due to the low level of decollimation, one can assume that,

$$\theta_1 + \theta_2 = 2\theta_B,$$

$$\sin\left(\frac{\theta_1-\theta_2}{2}\right) = \left(\frac{\theta_1-\theta_2}{2}\right).$$

Then, it is found that,

$$2\,t\left(\frac{\theta_1-\theta_2}{2}\right)\cos\theta_B = \lambda.$$

Finally, approaching the full width at half maximum equals to $\theta_1 - \theta_2$, it is found that,

$$B = \theta_1 - \theta_2 = \frac{\lambda}{t\,\cos\theta_B}.$$

Once more, and due to the relative nature of our model and the approximately constant t value for any reflection, we can keep the proportional factor as $B \propto \lambda / \cos\theta_B$.

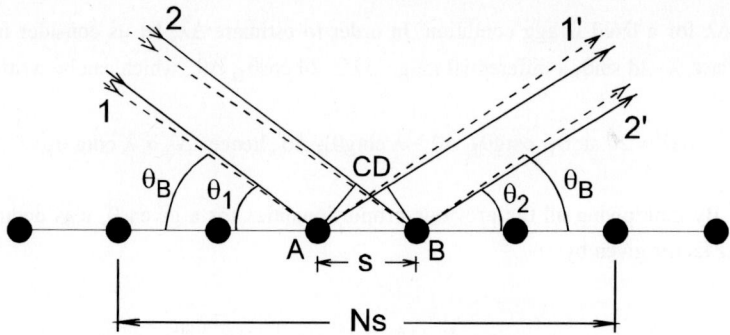

**Figure 2.8:** *Effect of de-collimation on maximum intensity of diffraction by bulk crystals.*

In order to estimate the $I_{max}$ value, let us consider the figure 2.8, where a plane with scattering centres spaced s is sketched. Let us considered a number N of the scattering centres which are irradiated by an incident beam with an angle $\theta_1 = \theta_B + \Delta\theta$, and by another incident beam with angle $\theta_2 = \theta_B - \Delta\theta$, $\theta_B$ being the Bragg angle for the given incident wavelength ($\lambda$). The path difference ($\delta$) between the beams scattered by the centres 1 and 2 is given by,

$$\delta = AD - CB = s \ \cos\theta_2 - s \ \cos\theta_1 = s \left[\cos(\theta_B + \Delta\theta) - \cos(\theta_B - \Delta\theta)\right].$$

Using trigonometric considerations and a first order approximation for $\sin\Delta\theta$, it is found that, $\delta = 2 \ s \ \Delta\theta \ \sin\theta_B$. Between the first and the last scattering centres there is a spacing Ns and therefore, they have a path difference $\delta_N = 2N \ s \ \Delta\theta \ \sin\theta_B$. For $\delta_N = \lambda$ the diffracted intensity is totally cancelled between the intermediate scattering centres. Hence, the limit value for $\Delta\theta$ which contributes to $I_{max}$, is given by $\lambda / 2Ns \ \sin\theta_B$, Therefore, neglecting the proportionalities, it can be written for each reflection that, $I_{max} \propto \lambda / \sin\theta_B$.

Together with the collimation effects, one must include in the Lorentz factor the spectral incident interval contributing to the integrated intensity. Although the estimation method proposed in this book uses a discretised treatment of the spectral distribution of the source, the interval for the continuous spectral decomposition will be set to an accuracy of 0.04Å on wavelength. Therefore, the previous considerations for a fixed $\lambda$ must be extended to a range $\Delta\lambda$. Hence, the effective intensity is proportional to the range $\Delta\lambda$ for a fixed Bragg condition. In order to estimate $\Delta\lambda$, let us consider from the Bragg law, $\lambda = 2d \sin\theta$, a differential range $\Delta\lambda = 2d \cos\theta_B \, \Delta\theta$, which can be written

$$\Delta\lambda = 2d \sin\theta_B \, \cot g\theta_B \, \Delta\theta = \lambda \cot g\theta_B \, \Delta\theta, \text{ hence, } \Delta\lambda \propto \lambda \cot g\,\theta_B.$$

By multiplying all the previous proportionalities for a given $\theta$, it is deduced the Lorentz factor given by:

$$LF = \frac{\lambda}{\cos\theta} \frac{\lambda}{\sin\theta} \, \lambda \cot g\theta \; = \; \frac{\lambda^3}{\sin^2\theta},$$

$\lambda/\cos\theta$ being the band width, $\lambda/\sin\theta$ the distribution maximum and $\lambda \cot g\theta$ the spectral interval.

## CHAPTER 3

## DEVELOPMENT OF THE COMPUTATIONAL PROCEDURES FOR THE SIMULATION AND INDEXING OF BACK-REFLECTION LAUE PATTERNS

Traditionally the orientation of a sample have been done appreciating possible symmetries from an experimental Laue-gram and checking the result by means of a new experimental pattern. For this work, the Wulff net and the Greninger chart (23.a) were employed to simplify the projection data analysis, and pioneer programs were developed to aid this time consuming task (24). But even for crystals of a kind most suited to this procedure, it was found necessary to take an average of 55 Laue-grams in order to unambiguously set the orientation of a sample (25).

Krahl-Urban, Butz & Preuss (26) were the first to produce computer simulations of Laue patterns for different orientations of some compounds (27). Cornelius (28) took into account the relative intensities of the Laue spots, following an idea of Preuss (29), which in general contains several approximations. Christiansen & Gerward (19) improved the simulation of Laue-grams by indexing the strongest spots produced by the characteristics lines of the material used as source.

Previous works on simulations allowed the indexing of planes by comparison between experimental and simulated Laue-grams. Several specific indexing methods have been proposed, which can be summarised as follows:

i) Huang, Christensen & Block (30), Lisbôa & Edwards (31), Ploc (32) and Riquet & Bonet (33) compared the angles between vectors of the reciprocal lattice with the ones of the experimental Laue-gram. The most efficient algorithm is the one developed by Riquet & Bonet, and latter applied by several authors (34, 35)

ii) Hart & Rietman (36) and Fewster (37) have proposed the identification of the Laue-grams using the zone axis concept. All crystals faces which are parallel to a given direction are said to belong to the same zone, and the given direction is called the zone axis. In the back reflection Laue projection, all the planes of a zone sketch an hyperbola in the pattern. This methods compare the angles of the zone axis measured in the experimental Laue-gram with the angles of the real lattice. By this method, one can try to reduce the computing time. Nevertheless, a large input data and the identification of the zone axis in the experimental Laue-gram (which is not always obvious) are necessary.

One can improve the calculation of the orientation by use of a least-squares method, following Laugier & Filhol (34). In this way, orientations with an error of 0.01° can be obtained by using a suitable detector and collimator (38).

In this chapter, the complete procedures for programming the equations involved in the simulation and indexing of Laue-grams are presented, using the methods proposed in references (35), (39) and (40). Several useful tips are provided in order to get the maximum advantage of the algorithms.

## 3.1 Geometrical simulation procedure

As explained in section 1.3.2, the adjacent planes that exist in a crystal structure are defined by the Miller indices h, k and l. In real space, these are the planes that go through the points defined by the vectors $a_1/h$, $a_2/k$ and $a_3/l$, where $a_1$, $a_2$ and $a_3$ are the vectors of the direct lattice.

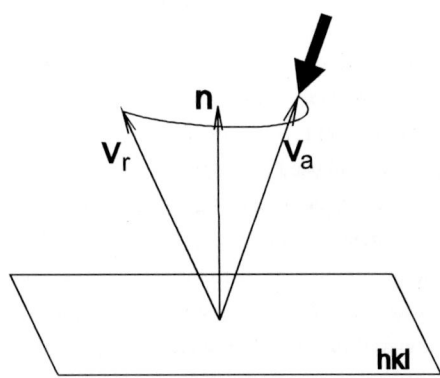

**Figure 3.1:** *Diagram of reflection by a (h k l) plane.*

As shown in figure 3.1, after obtaining a vector **n** normal to each of the planes (h k l) (see section 1.3.2), one can get the direction of any reflection ($V_r$) by a $\pi$ rotation of the antiparallel vector ($V_a$) to the incident direction, **n** being the rotation axis.

In order to get the cross point of the reflected beam on the detector, it is necessary to find the intersection between the line that contains the vector obtained after the rotation and the plane where the film is located. This point is a function of the distance and the angle between source, sample and film.

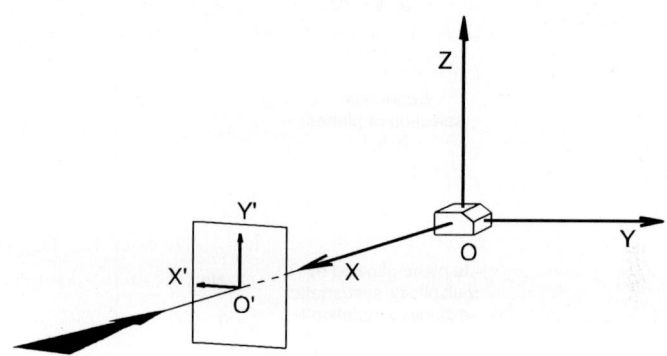

**Figure 3.2:** *Experimental geometry of the Laue back-reflection.*

The back-reflection Laue technique, described in section 2.1, is schematically shown in figure 3.2, where a reference system is fixed for the goniometer that contains the sample (O;X,Y,Z) and for the film (O';X',Y').

The flow chart of the algorithm for simulation of the spot positions of a Laue pattern is shown in figure 3.3, which follows the steps described below:

a) In the reference system (O;X,Y,Z), one assigns the direction of the incident radiation along the OX axis.

b) In this reference system, one estimates the coordinates of the direct lattice vectors $a_1$, $a_2$ and $a_3$, shown in figure 3.4, where $a_1$ is in the OX direction, and $a_2$ is in the plane XY;

$$a_1 = a_1 \, \mathbf{i}$$
$$a_2 = a_{2x} \, \mathbf{i} + a_{2y} \, \mathbf{j}$$
$$a_3 = a_{3x} \, \mathbf{i} + a_{3y} \, \mathbf{j} + a_{3z} \, \mathbf{k}$$

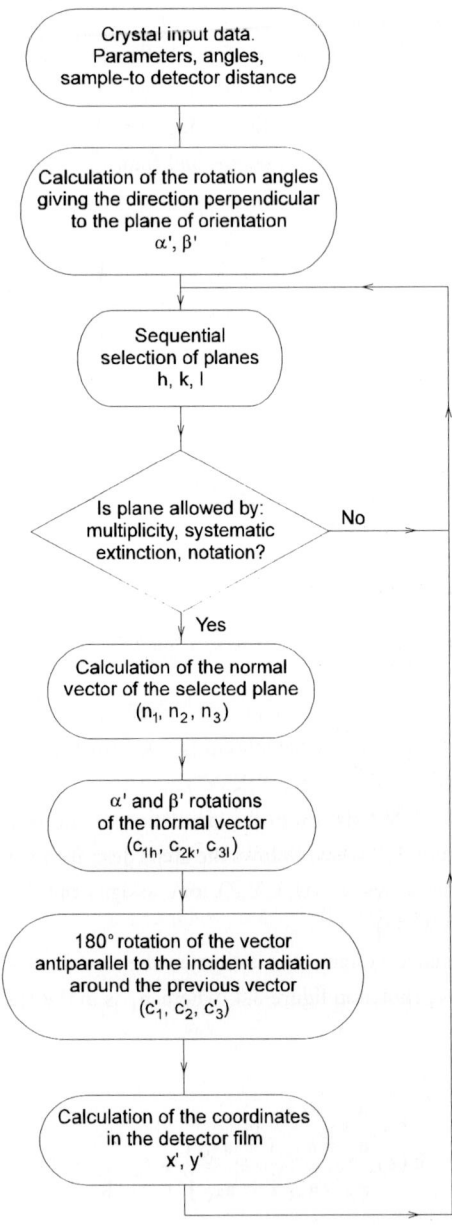

**Figure 3.3:** *Flow chart of the simulation procedure with the variables calculated at each stage shown.*

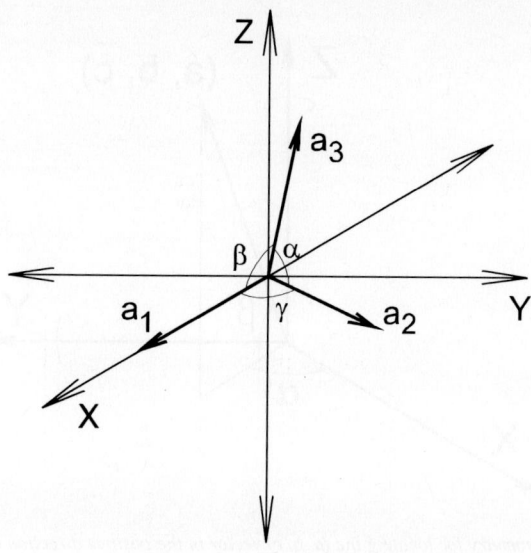

**Figure 3.4:** *General rules for the location of the vectors of a direct lattice.*

where **i**, **j** and **k** are the unit vectors of axes X, Y and Z respectively, and

$$a_{2x} = a_2 \cos\gamma$$
$$a_{2y} = a_2 \sin\gamma$$
$$a_{3x} = a_3 \cos\beta$$
$$a_{3y} = a_3 \left[(\cos\alpha - \cos\beta\cos\gamma)/\sin\gamma\right]$$
$$a_{3z} = a_3 \left\{\sin^2\beta - \left[(\cos\alpha - \cos\beta\cos\gamma)/\sin\gamma\right]^2\right\}^{1/2}$$

c) The normal vector **n** $(n_1, n_2, n_3)$ of a (h k l) plane will be given by,

$$n_1 = h\, a_{2y}\, a_{3z}$$
$$n_2 = -h\, a_{2x}\, a_{3z} + k\, a_1\, a_{3z}$$
$$n_3 = h\, (a_{2x}\, a_{3y} - a_{2y}\, a_{3x}) - k\, a_1\, a_{3y} + l\, a_1\, a_{2y}$$

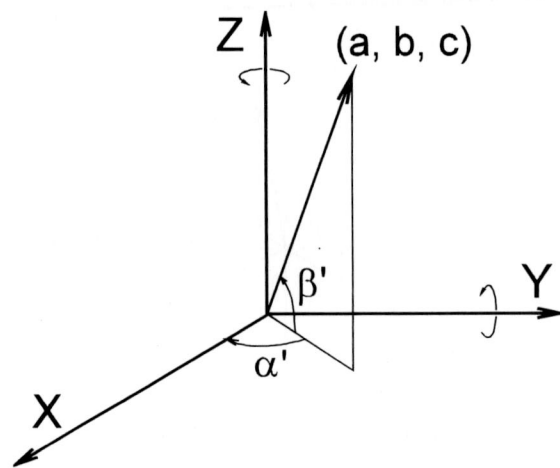

**Figure 3.5:** *Geometry for locating the (a, b, c) vector in the positive direction of the OX axis.*

d) In general, to locate any vector (a, b, c) in the positive direction of the OX axis, it is necessary to rotate this vector by the angles $\alpha'$ and $\beta'$ with respect to OY and OZ axes as shown in figure 3.5, where

$$\alpha' = \begin{cases} -\arctan(b/a) & \text{if } b \geq 0 \\ -\arctan(b/a) - \pi & \text{if } b < 0 \end{cases}$$

$$\beta' = -\arctan\left[c/(b^2 + a^2)^{1/2}\right],$$

giving the operator $\alpha'$,

$$\begin{pmatrix} \cos\alpha' & -\sin\alpha' & 0 \\ \sin\alpha' & \cos\alpha' & 0 \\ 0 & 0 & 1 \end{pmatrix},$$

and the operator $\beta'$,

$$\begin{pmatrix} \cos\beta' & 0 & -\sin\beta' \\ 0 & 1 & 0 \\ \sin\beta' & 0 & \cos\beta' \end{pmatrix}.$$

An oriented crystal in the plane $(h_1 k_1 l_1)$ will have its normal vector $(n_{11}, n_{12}, n_{13})$ in the direction of the positive OX axis. Consequently, all the planes $(h_i k_i l_i)$ might rotate $n_i$ by the $\alpha'$ and $\beta'$ angles, with values $(a = n_{11}, b = n_{12}, c = n_{13})$. The vectors obtained after these rotations are $(c_{1h}, c_{2k}, c_{3l})$.

e) For each plane (h k l), the antiparallel vector (1, 0, 0) to the incident radiation is rotated $\pi$ around the normal vector $(c_{1h}, c_{2k}, c_{3l})$ to get the reflection vector $\mathbf{V}_r$ of coordinates $(c'_{1h}, c'_{2k}, c'_{3l})$. The equations are (41):

$$
\begin{pmatrix} c'_{1h} \\ c'_{2k} \\ c'_{3l} \end{pmatrix} = \mathbf{B}^{-1} \begin{pmatrix} 1 & 0 & 0 \\ 0 & -1 & 0 \\ 0 & 0 & -1 \end{pmatrix} \mathbf{B} \begin{pmatrix} 1 \\ 0 \\ 0 \end{pmatrix},
$$

where

$$
\mathbf{B} = \begin{pmatrix} \left(\dfrac{c_{2k}^2}{c_{3l}c_{1h}}\right) + \left(\dfrac{c_{3l}}{c_{1h}}\right) & -c_{2k}/c_{3l} & -1 \\ 0 & 1 & -c_{2k}/c_{3l} \end{pmatrix}.
$$

f) Finally, the intersection point (x', y') in the reference system (O';X',Y'), between the projection of the former vectors $(c'_{1h}, c'_{2k}, c'_{3l})$ and the film is estimated. In the back-reflection geometry, for a given distance 's' between the sample and the film, the coordinates are given by $x' = -s\, c'_{2k}/c'_{1h}$ and $y' = s\, c'_{3l}/c'_{1h}$.

Here, we must take into account the following facts.

i) the equivalent Miller indices, should not be considered because the Laue pattern coordinates of each plane (nh nk nl) coincide in position. In the estimation of intensity (section 3.2.1) these planes will be considered as nth-order diffraction of the (h k l) plane.

ii) the destructive interference conditions that appear depending on the kind of Bravais lattice (42):

- body centered: h+k+l odd,
- all faces centered: h+k, k+l and l+h all odd,
- {1 0 0} centered: k+l odd,

- {0 1 0} centered: l+h odd,
- {0 0 1} centered: h+k odd.

In the model used for the estimation of intensities, these conditions will be dictated by the structure factor calculation (section 2.3.3).

## 3.2 Estimation of intensities procedure

### 3.2.1 Procedure for the estimation of intensities

The model proposed in this chapter estimates the relative intensity of each spot of an experimental Laue-gram and is included as a subroutine in the program described in the chapter 4. The special feature of this procedure is based on a spectral discretization of the continuous radiation. Taking into account the errors in computation, the time for calculation and the accuracy of the parameters used from International Tables for X-ray Crystallography (2), an optimal interval of 0.04Å has been employed for the continuous spectral discretization.

The following factors and effects have been incorporated in the present model:

1. The spectral sensitivity of the detector (section 2.3.2). The software herein included considers a conventional AgBr film detector.

2. The real spectral intensity of the incident radiation on the sample: the program is made for a conventional Mo cathodic source operating at its higher voltage below the appearance of its characteristics lines (20kV). No other elements are considered, therefore  the emitted radiation falls without any insertion between the source and the sample (section 2.3.1).

3. The temperature modifies the values of the form factor of any atom n ($f_n$) according to the expression $f_{nT} = f_n(\sin\theta / \lambda)\exp[-M(\sin\theta / \lambda, T, m_n, \Theta_D)]$, see section 2.3.4.

4. The anomalous dispersion (section 2.3.5) modifies each form factor to an effective value $f_{effn} = f_{nT} + \Delta f(\lambda / \lambda_{Kn})$, where $\Delta f$ is a function of the incident wavelength ($\lambda$) and the K emission wavelength of the n atom ($\lambda_{Kn}$).

5. The square of the modulus of the structure factor $|F|^2$ (section 2.3.3), which is proportional to the diffracted intensity.

6. The absorption factor AF (section 2.3.7).

7. The polarization factor PF (section 2.3.6).

8. The Lorentz factor LF (section 2.3.8).

It is necessary to take into account the effective reflection of the planes (nh nk nl) because several orders of diffraction may fall in the same position than the (h k l) diffraction, depending on the minimum wavelength value of the falling radiation.

The relative intensity of each spot is evaluated by multiplying all these factors. To develop the visual representation of the Laue-gram it is also necessary to put forward a minimum and a saturation level for the detection, taking into account the exposure time as main criterion. In the case of a film detector, the scaling between these two levels can be considered linear (section 3.2.2.c).

**Table 3.1:** *A priori data required for estimation of intensities.*

| | |
|---|---|
| 1 | Spectral intensity of incident radiation |
| | Spectral sensitivity of the detector |
| | Orientation to perform |
| | Distance from the sample to the center of the detector |
| 2 | Crystalline system of the sample |
| | Unit cell parameters and angles |
| | Number of atoms of the unit cell (N) |
| | Relative coordinates of the N atoms |
| | Debye temperature |
| 3 | Atomic weight of each atom of the material |
| | Wavelength of the absorption K edge of each atom |
| | Form factor of each kind of atom |
| | Linear absorption coefficient per weight of each atom |

Table 3.1 summarises the numerical values employed to perform the simulation. The first block indicates the experimental conditions required for the simulation. Then, the parameters of the second block are inherent for a particular material and will be stored in the computer memory. The third block are the data found in International Tables for X-ray Crystallography (2) and will also be stored permanently in the computer.

The flow chart of the algorithm proposed is shown in figure 3.6, and must be invoked for each (h k l) plane. The relationships of each factor and effect employed in the method have been arranged in the most logical way. The procedure shown in figure 3.6 specified the following steps for a selected orientation of the sample:

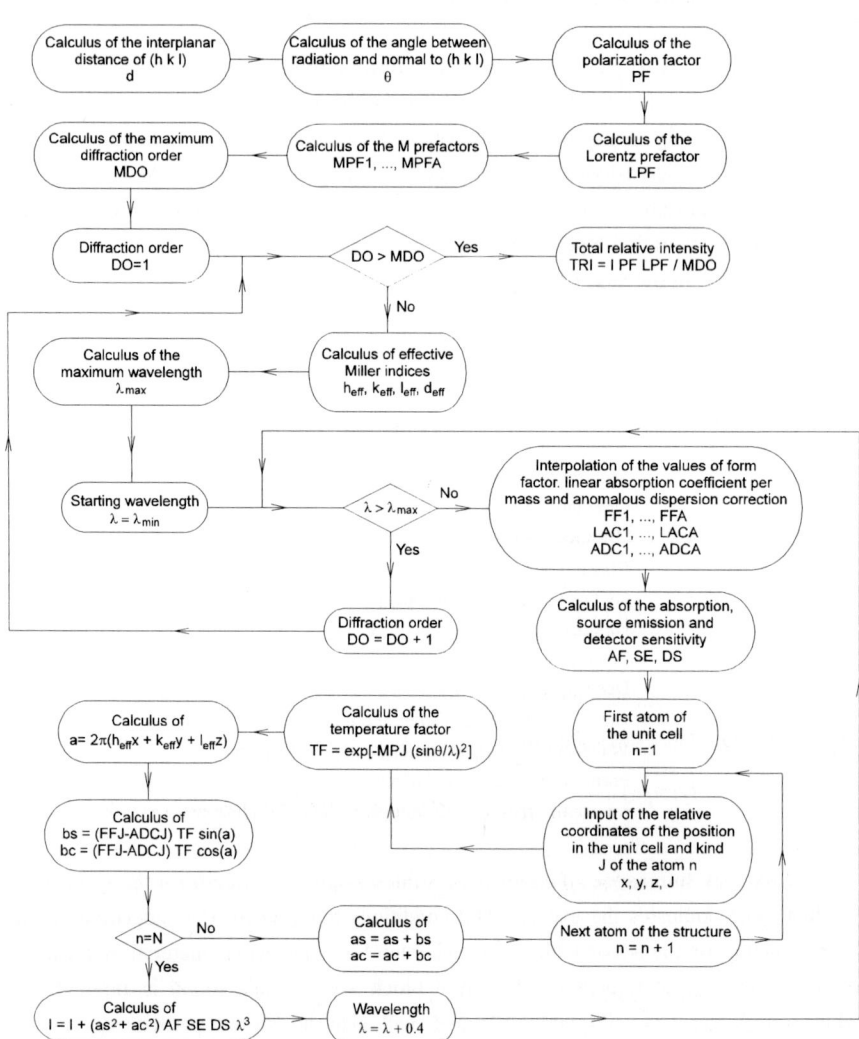

**Figure 3.6:** *Flow chart of the intensity estimation algorithm with the variables calculated at each stage shown.*

a) The interplanar distance (d) of the planes (h k l) is calculated in the usual way, as a function of the unit cell parameters (a, b, c, $\alpha$, $\beta$, $\gamma$) from:

$$\frac{1}{d^2} = \frac{1}{\Omega^2}(S_{11}h^2 + S_{22}k^2 + S_{33}l^2 + 2\,S_{12}h\,k + 2\,S_{23}k\,l + 2\,S_{13}h\,l)$$

where,
$$\begin{cases} \Omega = \text{volume of unit cell} \\ S_{11} = b^2c^2\sin^2\alpha \\ S_{22} = a^2c^2\sin^2\beta \\ S_{33} = a^2b^2\sin^2\gamma \\ S_{12} = a\,b\,c^2(\cos\alpha\,\cos\beta - \cos\gamma) \\ S_{23} = a^2b\,c\,(\cos\beta\,\cos\gamma - \cos\alpha) \\ S_{12} = a\,b^2c\,(\cos\gamma\,\cos\alpha - \cos\beta) \end{cases}$$

b) The angle $\theta$ between the antiparallel vector to the incident radiation and the normal to the plane (h k l) is calculated by means of the dot product.

c) The polarization factor $PF = (1 + \cos^2 2\theta)$ is calculated (section 2.3.6).

d) From the Lorentz factor $LF = \lambda^3 / \sin^2\theta$ (section 2.3.8), the Lorentz prefactor (LPF) is calculated as its angular contribution $LPF = 1/\sin^2\theta$.

e) From the expression for the argument (M) of the exponential of the temperature factor (section 2.3.4), the M prefactor (MPFJ) is calculated for each kind of atom J, from 1 to the number of kinds of atoms A present in the structure, which is given by,

$$M_J = (6h^2T/m_Jk_B\Theta_D^2)\left[\varphi(\Theta_D/T) + (\Theta_D/4T)\right](\sin\theta/\lambda)^2 = MPFJ\,(\sin\theta/\lambda)^2.$$

In order to reduce the running time of the simulation, the function $\varphi$, defined in section 2.3.4, should be numerically calculated for an accurate interval. Then their values should be stored in a file that will be loaded by the program.

f) The maximum diffraction order (MDO) of the (h k l) diffraction is a function of the minimum incident wavelength $\lambda_{min}$. If a cathodic tube is used, the value of $\lambda_{min}$ is hc/eV (see section 1.1), where h is the Planck constant, c the light velocity, e is the electron charge and V the applied voltage. Therefore the value of MDO is $2d \sin\theta / \lambda_{min}$, d being the interplanar distance.

g) For every diffraction order (DO) from 1 to MDO, the effective Miller indices ($h_{eff} = DO\, h$, $k_{eff} = DO\, k$, $l_{eff} = DO\, l$), the effective distance between planes ($d_{eff} = d/DO$) and the maximum wavelength that contributes to the diffraction ($\lambda_{max} = 2d \sin\theta / DO$) are set up.

h) With a discretization of 0.04Å, for every wavelength from $\lambda_{min}$ to the $\lambda_{max}$ of every DO, the following values are interpolated from International Tables for X-ray Crystallography (2):
   1. The form factors (FF) for all kinds of atoms that form the structure, FF1, ..., FFA;
   2. the mass absorption coefficients (LAC) defined as $\mu/\rho$, which is a function of $\lambda$, for all the kinds of atoms, LAC1, ..., LACA;
   3. the anomalous dispersion correction (ADC), which is a function of $\lambda/\lambda_K$, where $\lambda_K$ is the absorption wavelength of the kind of atom, ADC1, ..., ADCA.

i) The absorption factor (AF) of the material (section 2.3.7) is calculated:

$$AF = \left\{ \frac{\mu}{\rho}(\lambda) \left[ \cos(\pi/2 - \theta) + \frac{\cos(\pi/2 - \theta)}{\cos 2(\pi/2 - \theta)} \right] \right\}^{-1}.$$

Since relative intensities are considered, the mass absorption coefficient can be used. To calculate the $\mu/\rho$ values for the crystal, a weighted average of each element which constitutes the material is used with the weight fraction $w_J$, as

$$\left( \frac{\mu}{\rho} \right)_{material} = w_1 \left( \frac{\mu}{\rho} \right)_1 + ... + w_A \left( \frac{\mu}{\rho} \right)_A = w_1 LAC1 + ... + w_A LACA.$$

j) The relative intensity for the considered values of diffraction order (DO) and wavelength ($\lambda$), is calculated for each atom per unit cell, n=1 to N, N being the number of atoms per unit cell taken into account. Therefore, first the relative coordinates x, y and z of every atom are stored. Second the factors of the previous items (except the ones with geometrical dependence) are recorded, depending on the kind of atoms. Using the expression for the square of the structure factor (see section 2.3.3), the contribution for every $\lambda$ and diffraction order considered, $I_{\lambda,DO}$, is given by:

$$
\begin{aligned}
I_{\lambda,DO} = \Big[ \big( \{ & \ (FF1+ADC1)\exp[\ -MPF1(\sin\theta/\lambda)^2\ ] \\
& \times \cos 2\pi(h_{eff}x_1 + k_{eff}y_1 + l_{eff}z_1)\ \} + \ldots \\
& + \{ \ (FFN+ADCN)\exp[\ -MPFA(\sin\theta/\lambda)^2\ ] \\
& \times \cos 2\pi(h_{eff}x_N + k_{eff}y_N + l_{eff}z_N)\ \} \big)^2 \\
& + \big( \{ \ (FF1+ADC1)\exp[\ -MPF1(\sin\theta/\lambda)^2\ ] \\
& \times \sin 2\pi(h_{eff}x_1 + k_{eff}y_1 + l_{eff}z_1)\ \} + \ldots \\
& + \{ \ (FFN+ADCN)\exp[\ -MPFA(\sin\theta/\lambda)^2\ ] \\
& \times \sin 2\pi(h_{eff}x_N + k_{eff}y_N + l_{eff}z_N)\ \} \big)^2 \Big]\ AF\ SE\ SD\ \lambda^3,
\end{aligned}
$$

where $\lambda^3$ is the spectral contribution of the Lorentz factor.

k) The contributions of each DO and $\lambda$ are added. The factors with geometrical dependence applied are the Lorentz prefactor (LPF) and the polarization factor (PF). The effect of the multiple diffraction overlap (43) is taken into account using the average of all the DO involved as a first approach. Therefore, the total contribution is divided by MDO. In this way, the total relative intensity (TRI) for the spot labelled (h k l), $I_{hkl}$, is

$$
I_{hkl} = MDO^{-1}\ PF\ LPF \sum_{\lambda,DO} I_{\lambda,DO}.
$$

### 3.2.2 Results and comments about the estimation of intensities

#### 3.2.2.a Comparison between the simulation of back-reflection patterns with and without estimation of intensities

Figure 3.7a shows an experimental Laue-gram of a rhombohedral $LiNbO_3$ single crystal (appendix A2.7) in the (0 0 0 1) orientation (hexagonal notation, see 23.b). The corresponding simulated patterns without and with intensity estimations are shown in figures 3.7b and 3.7c respectively.

Laue-gram shown in figure 3.7c has been plotted with six different spot sizes, which represent six intensity levels, linearly scaled from the best fitting of the detection (DL) and saturation (SL) levels, in % from the highest estimated intensity (see section 3.2.1).

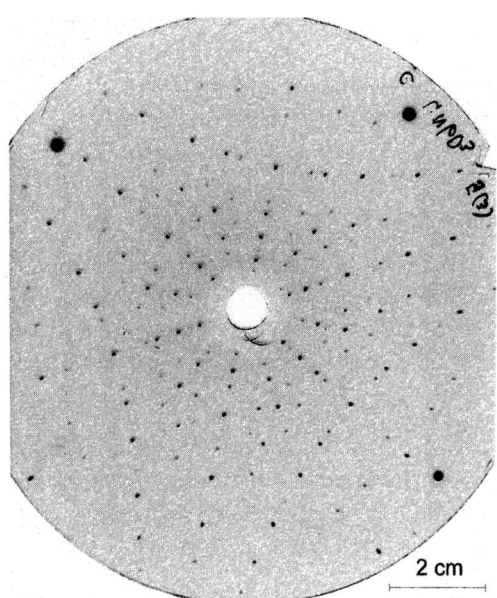

2 cm

**Figure 3.7a:** *Experimental back-reflection Laue-gram for $LiNbO_3$: plane (0 0 0 3), Mo tube V=20kV, I=40mA, exposure time 150 min., Kodak AX film, sample-to-detector distance 3.05 cm.*

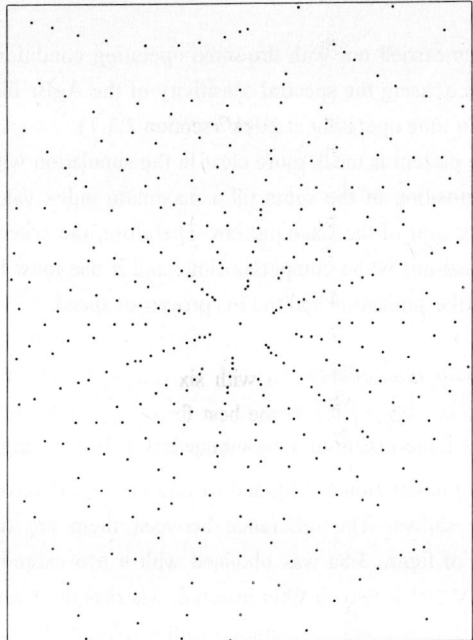

**Figure 3.7b:** *Simulated back-reflection Laue-gram for LiNbO₃, plane (0 0 0 3), maximum Miller index 30, sample-to-detector distance 3.05 cm.*

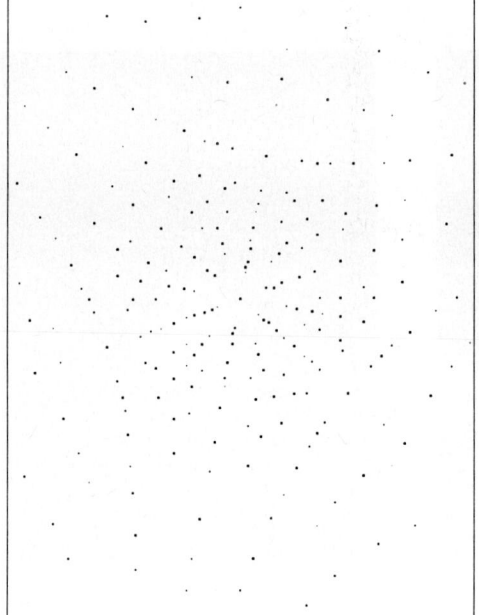

**Figure 3.7c:** *Simulated back-reflection Laue-gram for LiNbO₃, plane (0 0 0 3), maximum Miller index 40, detection level 2%, saturation level 45%, sample-to-detector distance 3.05 cm.*

....

The experimental laue-gram was carried out with the same operating conditions employed in the simulation procedure i.e. using the spectral sensitivity of the AgBr film (section 2.3.2) and the emission of a Mo tube operating at 20kV (section 2.3.1).

Although the recognition of the pattern is much more clear in the simulation with intensity estimations, the geometrical position of the spots till a maximum index value (see section 1.5) is enough to index any spot of the Laue pattern. Therefore, the criteria to work with or without intensity estimations is the computing time, and it use must be decided depending on the objectives of the simulation and the PC processor speed.

### 3.2.2.b Filtering and characteristic lines effects

In figure 3.8, two experimental Laue-grams of a rhombohedral $LiNbO_3$ single crystal (appendix A2.7) in the ($3\ 0\ \overline{3}\ 0$) orientation (23.b), and carried out in a Polaroid 57 intensifying screen film type, are shown. The difference between them are the experimental condition: the laue-gram of figure 3.8a was obtained with a Mo cathodic source operating with a voltage of 30kV and with a Zr filter inserted, whereas the Laue-gram in figure 3.8b was obtained with the same source operating with a voltage of 20kV and removing the filter.

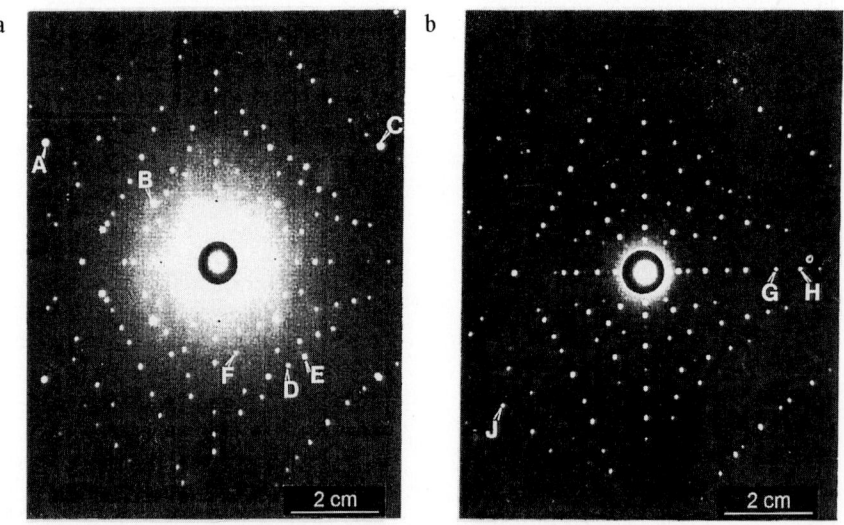

**Figure 3.8:** *Experimental back-reflection Laue-grams for a $LiNbO_3$ crystal: orientation ($3\ 0\ \overline{3}\ 0$), Mo X-ray tube, Polaroid 57 film, exposure time 20 min., sample-to-detector distance 4.1cm., a) V=30kV, I=40mA, Zr filter, b) V=20kV, I=40mA.*

To understand this experimental result, the Mo emission spectra, shown in figure 1.2 of the section 1.1, must be taken into account. It can be seen that, employing a voltage of 30kV, the Mo K characteristic lines are excited, whereas 20kV keeps the emission below the threshold of its excitation. Furthermore, the Zr insertion affects all the spectral distribution, accordingly with the spectral dependence of its absorption coefficient.

The following consequences can be obtained from the observation of the Laue-grams shown in figure 3.8:

i) The presence of spots of high intensity in figure 3.8a, such as the ones labelled A ($12\ \bar{3}\ \bar{9}\ 12$), B ($13\ \bar{2}\ \bar{11}\ 6$) and C ($4\ \bar{1}\ \bar{3}\ \bar{4}$), are due to the contribution to the diffraction intensities of the wavelengths of the characteristic K lines of the Mo ($\lambda_{K_\alpha} = 0.711$Å, $\lambda_{K_\beta} = 0.632$ Å), even when the Zr filter is used. These intense spots increase its size when the voltage applied to the tube is increased and an intensifying film is employed (section 1.2). These spots are the ones considered for the simulation and indexing work by some proposed procedures (19). In that cases, the applied voltage reaches more than 50kV and the study is limited to the diffraction conditions promoted by these spots. They are the only ones that appear in the Laue patterns. However, this reduction in the number of detected diffraction conditions easily promotes doubts during the identification between orientations with the same symmetries.

ii) It can be seen that spots D ($9\ 3\ \bar{12}\ \bar{6}$), E ($10\ 3\ \bar{13}\ \bar{8}$) and F ($10\ 3\ \bar{13}\ \bar{2}$) are only seen in Laue-gram shown in figure 3.8a. However, the reason for the absence of D, E and F are different. The spot F is absent in figure 3.8b because the $\lambda_{min}$ of the radiation is shifted to higher value, due to which this diffraction conditions are impossible. On the other hand, D and E are arising in figure 3.8a from the K characteristic wavelengths of the radiation (Mo) falling on the sample. For figure 3.8b these characteristic wavelengths are absent and the continuous radiation falling on the sample is less than the required one to reach the detection level during the exposure time employed.

iii) The opposite effect is also seen. Spots present in the Laue-gram of figure 3.8b cannot be seen in the Laue-gram of figure 3.8a, such as G ($3\ 0\ \bar{3}\ \bar{3}$), H ($7\ 0\ \bar{7}\ 8$) and J ($5\ 2\ \bar{7}\ 6$). In this case, the detection level is not reached in the experiment shown in figure 3.8a, because the Zr filtering reduces the incident intensities with the wavelengths that mainly contributes to the diffraction condition.

iv) Finally, it must be pointed out that there is an increase in the diffuse scattering with lower wavelength values of incident radiation, as discussed in section 2.3.4.

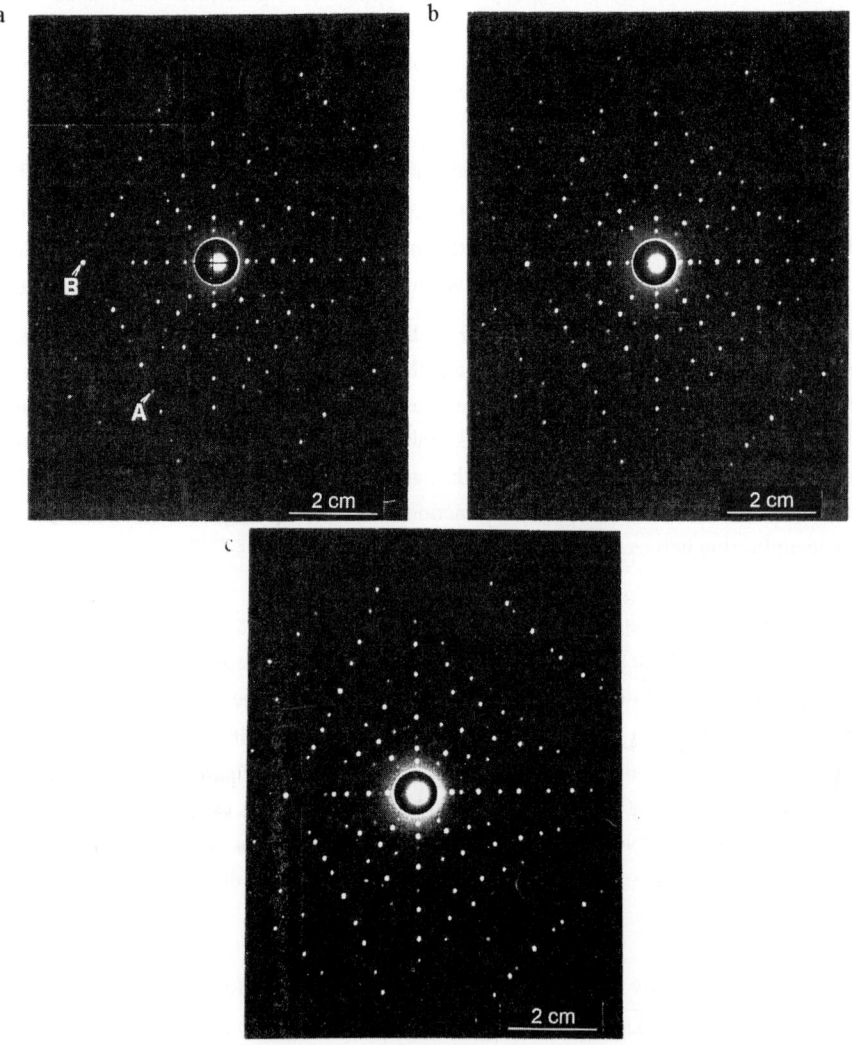

**Figure 3.9:** *Experimental back-reflection Laue-grams for a LiNbO₃ crystal: orientation (3 0 3̄ 0), Mo X-ray tube, V=20kV, I=40mA, Polaroid 57 film, sample-to-detector distance 4.1cm., exposure time a) 10 min., b) 15 min., c) 20 min.*

### 3.2.2.c Exposure time effects

The effects of the exposure time on the experimental back-reflection Laue patterns can be seen in the series of patterns in figure 3.9 for a $LiNbO_3$ single crystal in the ($3\ 0\ \overline{3}\ 0$) orientation (see appendix A2.7) carried out with an exposure time of 10, 15 and 20 minutes.

The points to be noted from these figures are:

i) The linearity in the intensity increment of the spots with increasing exposure time, it can be seen in spots which are initially far from the saturation level, such as the spot labelled A ($7\ 3\ \overline{10}\ 4$).

ii) The low variation in the intensity of spots initially near the saturation level as B ($1\ 0\ \overline{1}\ 1$). The spot size increases once the saturation level is reached, as expected for the intensifying film employed (see section 1.2).

### 3.2.2.d Film detector effects

Figure 3.10 shows two experimental Laue-grams of a rhombohedral $LiNbO_3$ single crystal in the ($3\ 0\ \overline{3}\ 0$) orientation (23.b) carried out with two kind of films: an AgBr type (Kodak AX), and an intensifying type (Polaroid 57).

The spectral sensitivity is different for each detector, due to the material and the technological process involved (see section 1.2). Therefore, between the Laue-grams shown in figure 3.10a and 3.10b, some opposite intensity relations can be seen, such as the one between spots labelled A ($8\ \overline{1}\ \overline{7}\ 6$) and B ($3\ 0\ \overline{3}\ \overline{3}$) in figure 3.10a. While in the Laue-gram a) the spot A is more intense than B, in the Laue-gram b) the spot B is more intense than A.

Additionally, the difference between the detection and saturation levels (see section 3.2.1) is wider in the AgBr type of film.

a

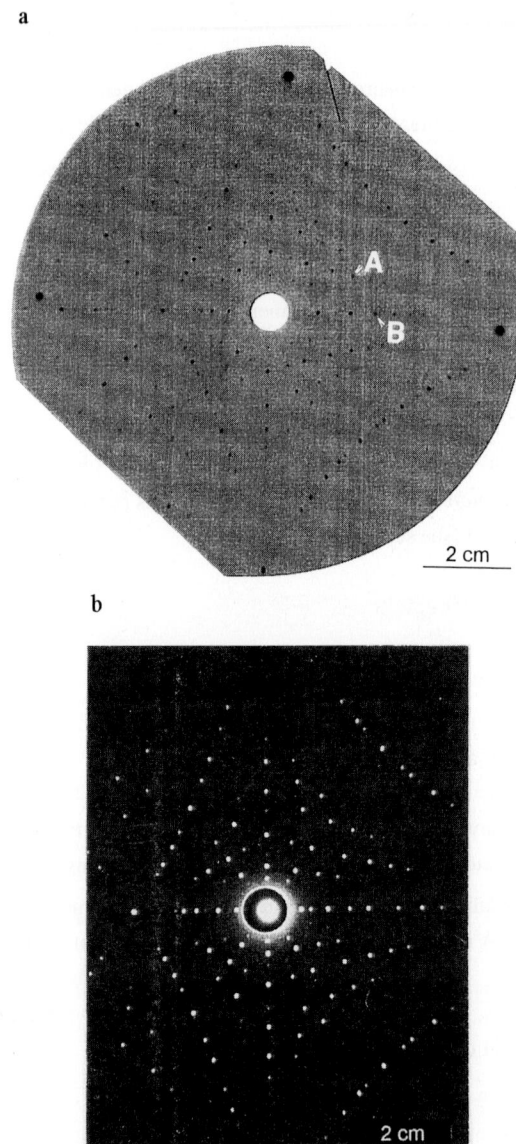

b

**Figure 3.10:** *Experimental back-reflection Laue-grams for a LiNbO₃ crystal: orientation (3 0 $\overline{3}$ 0), Mo X-ray tube, V=20kV, I=40mA; a) sample-to-detector distance 3cm., exposure time 150 min., Kodak AX film; b) sample-to-detector distance 4.1cm, exposure time 20 min., Polaroid 57 film.*

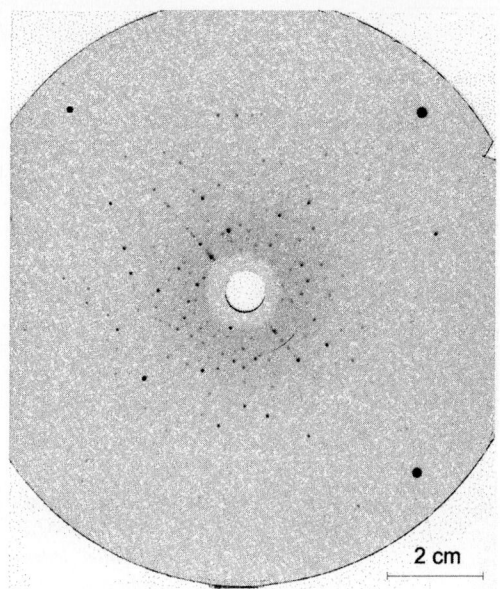

2 cm

**Figure 3.11a:** *Experimental back-reflection Laue-gram for: $Bi_{12}GeO_{20}$ plane $(\bar{2}\,\bar{1}\,\bar{1})$, Mo tube V=20kV, I=40mA, exposure time 180 min., Kodak AX film, sample-to-detector distance 3.15 cm.*

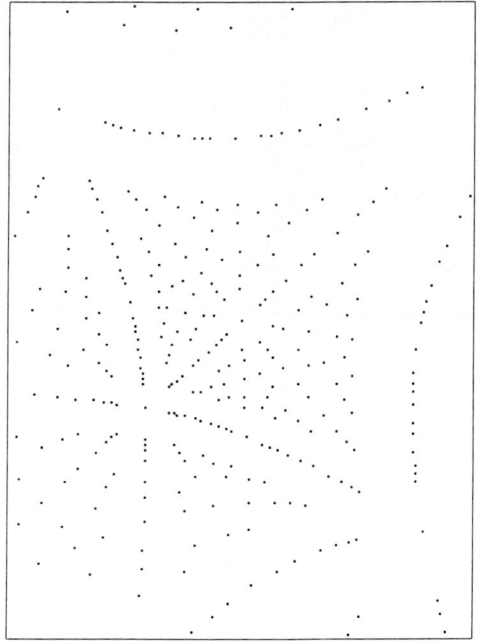

**Figure 3.11b:** *Simulated back-reflection Laue-gram for $Bi_{12}GeO_{20}$, plane $(\bar{2}\,\bar{1}\,\bar{1})$ or $(2\,1\,1)$, maximum Miller index 8, sample-to-detector distance 3.15 cm.*

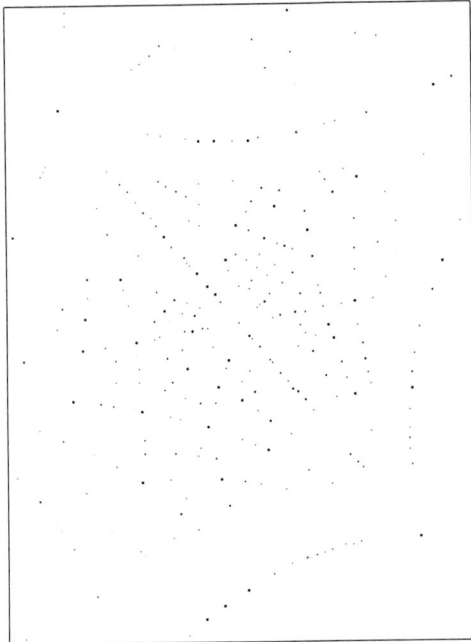

**Figure 3.11c:** *Simulated back-reflection Laue-gram for $Bi_{12}GeO_{20}$ plane $(2\,\overline{1}\,\overline{1})$, maximun Miller index 15, detection level 1%, saturation level 20%, sample-to-detector distance 3.15 cm.*

**Figure 3.11d:** *Simulated back-reflection Laue-gram for $Bi_{12}GeO_{20}$, plane $(2\,1\,1)$, maximum Miller index 15, detection level 1%, saturation level 20%, sample-to-detector distance 3.15 cm.*

### 3.2.2.e Distinction of anisotropic orientations

In order to carry out unambiguous indexing of anisotropic orientations, the intensity estimation becomes mandatory.

As an example, an experimental Laue-gram of one $Bi_{12}GeO_{20}$ cubic crystal (see appendix A2.1) is shown in figure 3.11a. This structure belongs to the $I23$ space group (see appendix A1) and, therefore, the ( 2 1 1) orientation is not equivalent to the ($\bar{2}\,\bar{1}\,\bar{1}$).

The geometrical simulation of the ( 2 1 1) and the ($\bar{2}\,\bar{1}\,\bar{1}$) orientations give rise to the same result (figure 3.11b). However, when the estimation of intensities is done, the simulated Laue-grams shown in figure 3.11c ($\bar{2}\,\bar{1}\,\bar{1}$) and 3.11d ( 2 1 1) allows us to unambiguously recognise the experimental Laue-gram as the ($\bar{2}\,\bar{1}\,\bar{1}$) orientation.

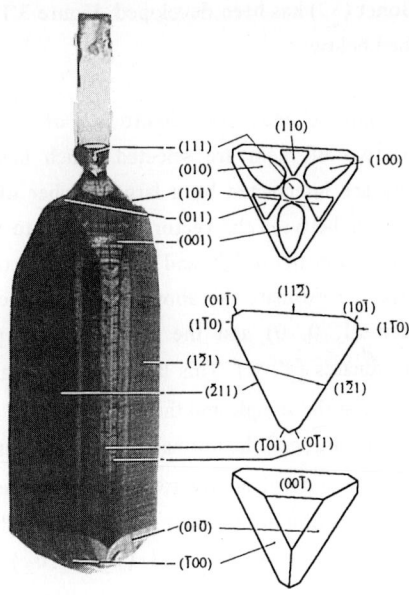

**Figure 3.12:** *Crystal of $Bi_{12}GeO_{20}$, grown by Czochralski technique pulling along the (1 1 1) direction, showing the orientation of the facets.*

The typical $Bi_{12}GeO_{20}$ single crystal shown in figure 3.12 has several facets (44). The unambiguous orientation of all these facets have been obtained using the present program, starting from an initial arbitrary direction which is determined from an initial arbitrary experimental pattern. The agreement of this results have been experimentally verified. This procedure is general and also useful to obtain oriented wafers from bulk single crystals or to get oriented seeds from unseeded grown single crystals.

## 3.3 Indexing method

### 3.3.1 Indexing procedure

For indexing of any back-reflection Laue pattern, an algorithm similar to the one proposed by Riquet & Bonet (33) has been developed. Figure 3.13 shows the flow chart of the procedure, described below:

a) From the coordinates x′ and y′ (see figure 3.2) of several experimental spots that are to be indexed, two of them are selected which have low Miller indices, recognisable because they are intersected by a large number of hyperbolas. For each spot, one estimates the angles between the vector antiparallel to the X-ray beam, which is fixed at (1, 0, 0) as depicted in figure 3.2, and the normal vector to each plane which reflects the selected spots. To estimate the above angles, one calculates the bisecting vector (a, b, c) between (1, 0, 0) and the vector whose projection creates the experimental spot of coordinates (x′, y′). This latter vector is given by $(1, -x'/s, y'/s)$, where s is the distance between the sample and the detector.

Generally, we call $(a_n, b_n, c_n)$ the normal vectors originated from the n selected experimental spots, where n=1 and n=2 are the two chosen as fundamental ones (a priori lower Miller indices). By means of a dot product, one estimates the angle $\chi$ between the normal vectors of the planes $(a_1, b_1, c_1)$ and $(a_2, b_2, c_2)$ that reflect the two fundamental spots.

b) The next step is to calculate the angle $\xi$ between each pair of planes $(h_1\ k_1\ l_1)$ and $(h_2\ k_2\ l_2)$ that are going to be tested as assignable to $(a_1, b_1, c_1)$ and $(a_2, b_2, c_2)$. When the difference between the angles $\chi$ and $\xi$ is bigger than a chosen maximum

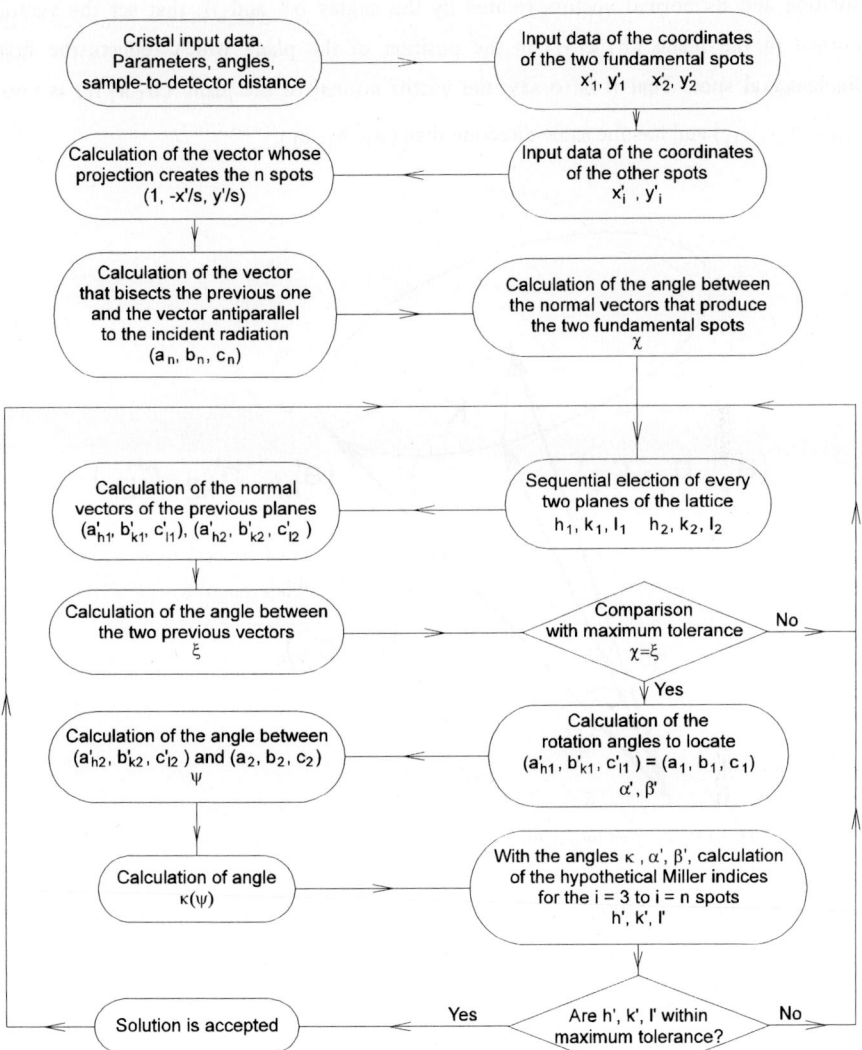

**Figure 3.13:** *Flow chart of the indexing program.*

angular tolerance, the pair of planes are discarded and a new pair is tested. When the difference is less than a maximum angular tolerance, these planes are selected as possible solution and its normal vectors rotated by the angles $\alpha'$ and $\beta'$ that set the vector normal to the plane $(h_1\,k_1\,l_1)$ in the position of the plane which reflects the first fundamental spot. That it is to say, the vector normal to the plane $(h_1\,k_1\,l_1)$ is now $(a'_{h1},\,b'_{k1},\,c'_{l1})$ and has the same direction than $(a_1,\,b_1,\,c_1)$.

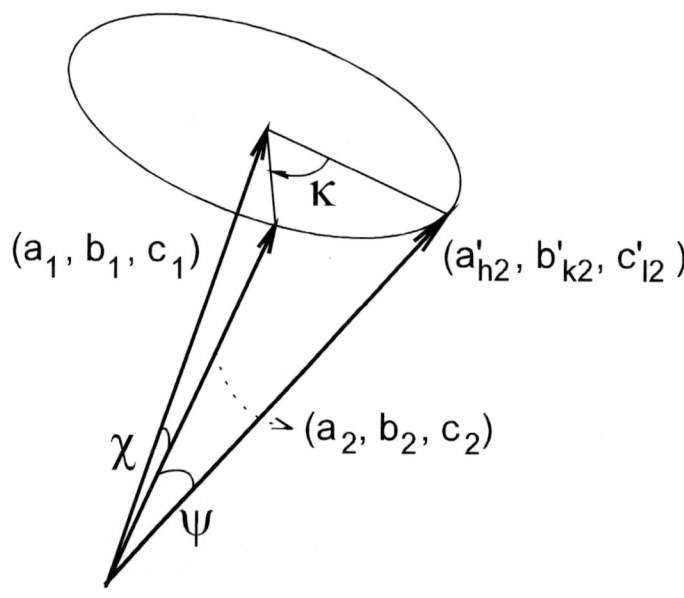

**Figure 3.14:** *Diagram for locating the vector $(a'_{h2},\,b'_{k2},\,c'_{l2})$ in position $(a_2,\,b_2,\,c_2)$.*

c) With the rotation axis as the vector in the position $(a_1,\,b_1,\,c_1)$, the vector normal to the second plane $(a'_{h2},\,b'_{k2},\,c'_{l2})$ is rotated in order to locate it in the position of the Laue model $(a_2,\,b_2,\,c_2)$, as shown in figure 3.14. In this way, if the angle $\psi$ between $(a_2,\,b_2,\,c_2)$ and $(a'_{h2},\,b'_{k2},\,c'_{l2})$ is known, the angle of rotation $\kappa$ is given by $\kappa\;=\;2\arcsin\left\{\sin\psi\,/\,2\sin[(\pi-\psi)/2]\sin\theta\right\}$, and one operates as follows:

$$\begin{pmatrix} a_2 \\ b_2 \\ c_2 \end{pmatrix} = \begin{pmatrix} \cos\Theta_2 & -\sin\Theta_2 & 0 \\ \sin\Theta_2 & \cos\Theta_2 & 0 \\ 0 & 0 & 1 \end{pmatrix} \begin{pmatrix} \cos(-\Theta_1) & 0 & \sin(-\Theta_1) \\ 0 & 1 & 0 \\ -\sin(-\Theta_1) & 0 & \cos(-\Theta_1) \end{pmatrix} \begin{pmatrix} 1 & 0 & 0 \\ 0 & \cos\kappa & -\sin\kappa \\ 0 & \sin\kappa & \cos\kappa \end{pmatrix}$$

$$\times \begin{pmatrix} \cos\Theta_1 & 0 & \sin\Theta_1 \\ 0 & 1 & 0 \\ -\sin\Theta_1 & 0 & \cos\Theta_1 \end{pmatrix} \begin{pmatrix} \cos(-\Theta_2) & -\sin(-\Theta_2) & 0 \\ \sin(-\Theta_2) & \cos(-\Theta_2) & 0 \\ 0 & 0 & 1 \end{pmatrix} \begin{pmatrix} a'_{h2} \\ b'_{k2} \\ c'_{l2} \end{pmatrix}$$

where

$$\Theta_1 = \arctan(b_1/a_1),$$
$$\Theta_2 = \arctan\left[c_1/(a_1+b_1)^{1/2}\right].$$

d) Finally, for the remaining vectors ($a_i$, $b_i$, $c_i$) with i=3 to i=n, one goes the opposite way, that is, one makes a rotation of $-\kappa$ with respect to ($a_1$, $b_1$, $c_1$) and of $-\beta'$ and $-\alpha$ with respect to the axes OY and OZ. After this rotation, one would get the hypothetical Miller indices of these vectors from the expression for the normal vector to a plane (h k l) mentioned in section 3.2.1 point c). The solution is valid if these indices are within the maximum Miller index tolerance chosen. If not, the solution is neglected and, as the process is iterative, a new pair of planes is selected and the process is repeated from step 2.

*3.3.2 Indexing result*

Figure 3.15a shows an experimental Laue-gram of an unknown orientation of a LiNbO$_3$ single crystal. To proceed with the indexing of the Laue-gram, one selects several spots (1 to 6), considering spots 1 and 2 as fundamental, according with the rules discussed in section 3.3.1. Then, by introducing into the algorithm the indexing data shown in table 3.2, several solutions will be given, as shown in table 3.3. Finally by use of the simulation algorithm discussed in sections 3.1 and 3.2, the simulated Laue-gram is obtained, as shown in figure 3.15b. This simulated Laue-gram corresponds to all indexing solutions presented in table 3.3 for the symmetry involved. Also in table 3.3 are indexed the spots A and B of figure 3.15a, which were not used as the input data and which support the general rule that any experimental spot can be indexed.

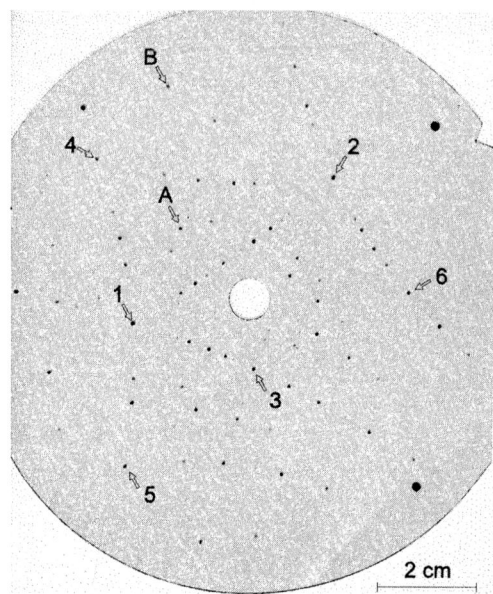

**Figure 3.15a:** *Experimental back-reflection Laue-gram for LiNbO₃: Mo tube V=20kV, I=40mA, exposure time 150 min., Kodak AX film, sample-to-detector distance 3.05 cm.*

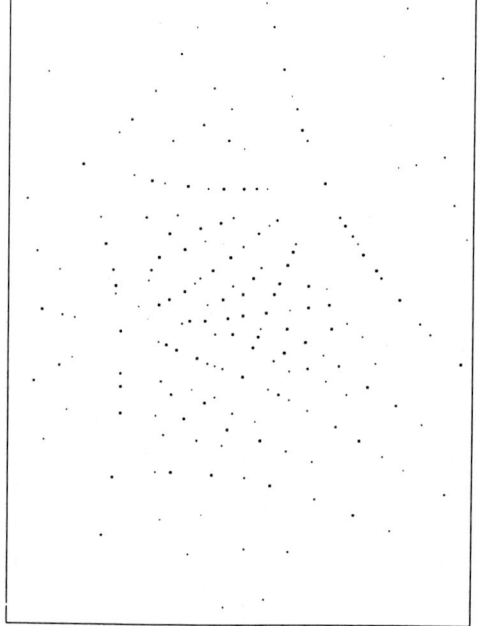

**Figure 3.15b:** *Simulated back-reflection Laue-gram for LiNbO₃. Solution 1 in table 3.3; maximum Miller index 40, detection level 10%, saturation level 50%, sample-to-detector distance 3.05 cm.*

**Table 3.2:** *Inputs for the indexing of the Laue-gram in figure 3.15.*

|  |  |  |
|---|---|---|
| Maximum Miller index | | 6 |
| Sample-to-detector distance (cm) | | 3.05 |
| Maximum angular tolerance (°) | | 0.5 |
| Maximum index tolerance | | 0.1 |
|  | x | y |
| Coordinates of fundamental spots (cm) | | |
| Spot 1 | -2.350 (25) | -0.500 (25) |
| Spot 2 | 1.650 (25) | 2.400 (25) |
| Coordinates of the other spots (cm) | | |
| Spot 3 | 0.100 (25) | -1.400 (25) |
| Spot 4 | -3.150 (25) | 2.700 (25) |
| Spot 5 | -2.500 (25) | -3.300 (25) |
| Spot 6 | 3.150 (25) | 0.150 (25) |

**Table 3.3:** *Indexing solutions for the Laue-gram in figure 3.15 with the inputs of table 3.2.*

|  | Solution 1 | Solution 2 | Solution 3 | Solution 4 |
|---|---|---|---|---|
| 1 | $0\ 0\ 0\ \bar{3}$ | $0\ 0\ 0\ \bar{3}$ | $0\ 0\ 0\ \bar{3}$ | $0\ 0\ 0\ \bar{3}$ |
| 2 | $\bar{1}\ 0\ 1\ \bar{4}$ | $0\ 1\ \bar{1}\ \bar{4}$ | $1\ \bar{1}\ 0\ \bar{4}$ | $\bar{1}\ 1\ 0\ 4$ |
| 3 | $\bar{1}\ 1\ 0\ \bar{8}$ | $1\ 0\ \bar{1}\ \bar{8}$ | $0\ 1\ \bar{1}\ \bar{8}$ | $\bar{1}\ 0\ 1\ 8$ |
| 4 | $0\ \bar{1}\ 1\ \bar{8}$ | $\bar{1}\ 1\ 0\ \bar{8}$ | $1\ 0\ \bar{1}\ \bar{8}$ | $0\ 1\ \bar{1}\ 8$ |
| 5 | $0\ 1\ \bar{1}\ \overline{10}$ | $1\ \bar{1}\ 0\ \overline{10}$ | $\bar{1}\ 0\ 1\ \overline{10}$ | $0\ \bar{1}\ 1\ 10$ |
| 6 | $\bar{2}\ 1\ 1\ \bar{6}$ | $1\ 1\ \bar{2}\ \bar{6}$ | $1\ \bar{2}\ 1\ \bar{6}$ | $\bar{2}\ 1\ 1\ 6$ |
| A | $\bar{1}\ 1\ 2\ \overline{18}$ | $1\ 2\ \bar{1}\ \overline{18}$ | $2\ \bar{1}\ \bar{1}\ 18$ | $\bar{1}\ 2\ 1\ 18$ |
| B | $\bar{1}\ 2\ 3\ \overline{14}$ | $2\ 3\ \bar{1}\ 14$ | $3\ \bar{1}\ \bar{2}\ 14$ | $1\ \bar{3}\ \bar{2}\ 14$ |

|  | Solution 5 | Solution 6 |
|---|---|---|
| 1 | $0\ 0\ 0\ 3$ | $0\ 0\ 0\ 3$ |
| 2 | $0\ \bar{1}\ 1\ 4$ | $1\ 0\ \bar{1}\ 4$ |
| 3 | $1\ \bar{1}\ 0\ 8$ | $0\ 1\ \bar{1}\ 8$ |
| 4 | $\bar{1}\ 0\ 1\ 8$ | $1\ \bar{1}\ 0\ 8$ |
| 5 | $1\ 0\ \bar{1}\ 10$ | $\bar{1}\ 1\ 0\ 10$ |
| 6 | $1\ \bar{2}\ 1\ 6$ | $1\ 1\ \bar{2}\ 6$ |
| A | $\bar{1}\ \bar{1}\ 2\ 18$ | $2\ \bar{1}\ \bar{1}\ 18$ |
| B | $\bar{2}\ 1\ 3\ 14$ | $3\ \bar{2}\ \bar{1}\ 14$ |

In summary, the simulation model will give the solutions used in the algorithm for indexing and, by comparing experimental and simulated Laue-grams, one can index all the experimental spots and consequently the orientation of the sample. With the data, one can easily obtain the rotation angles needed to obtain any specific orientation.

### 3.3.3 Strategy for the indexing work

With the use of indexing and simulation programs, the traditionally hard orientation work becomes a simple task. Just one experimental Laue-gram is enough to orient a sample in all the required planes, if a proper compatible equipment is available (see section 2.1).

By using the indexing procedure discussed in section 3.3.1, and as it was shown in the example of section 3.3.2, the last step to select the correct orientation of an experimental Laue-gram is the visual comparison with the simulated patterns proposed as solution.

In this way, once a Laue pattern is recognised, the rotation angles to get any desired orientation can be obtained. Coming back to the initial position, any other plane can be obtained without any more X-ray experiment, and so on with all the required orientations.

From the survey of the indexing procedure of section 3.3.1, three elements determine the number of suggested solutions:

i) The tolerance values for angle and index:

Obviously, the higher the tolerance values, the higher number of solutions suggested. However, it cannot be extremely low, because of the propagation of errors introduced by the accuracy of the measurements, mainly the sample-detector distance (s) as mentioned in section 2.1.

ii) The maximum Miller index (mmi):

The number of pair of planes to be tested in the step b) of the algorithm described in section 3.3.1 is given by the value of the maximum Miller index (mmi). The combinations of planes which can be designable employing index values within the range from −mmi to mmi, will be tested.

The number of combinations and, therefore, the computing time and number of possible solutions, increases dramatically with the higher maximum Miller index value.

Therefore, it is important to chose fundamental spots of low Miller index. In general, the main selective criterion must not be to consider a fundamental spot as the one of higher intensity. In this sense, the longer the sample-detector distance, the lower number of low Miller index spots appear in the pattern, because less projections of the diffraction directions intersect with a detector of standard dimensions.

Finally, it must be pointed out that, when hexagonal notation is being used for a rhombohedral structure (23.b), the Miller indices of the spots are higher compared with other cases.

iii) The number of experimental points to be tested in the algorithm:

Clearly, the higher the number of spots, the higher number of possible solutions discarded. However, it is not advisable to start the guess with a high number of spots, because of measurement errors (mainly the sample-detector distance error), which even may discard the correct solution.

From the comments above, it is suggested to set the experimental sample-detector distance between 3 and 4 centimetres, and to start the indexing work selecting five experimental spots in the algorithm, including the two fundamental ones. The maximum hyperbolas intersection criterion should be employed. At least two spots with Miller indices lower than 3 ought to be found. A first value of 1 degree for angular tolerance and 0.1 for index tolerance are advisable.

If few solutions are proposed, the first step should be to increase the tolerance values to the ones which offer a number of solutions which could be checked through the visualisation of their simulations. If we are working with a crystal with a high number of symmetries, a lot of solutions will be suggested, which belong to the same simulated pattern. It will be necessary to visualise just one of them

If a large number of solutions are proposed, the number of experimental spots should be increased. If a correct solution is not found, the maximum Miller index ought to be increased, which would mean that the fundamental spots have higher values than the expected ones. In order not to have too many solutions, the tolerance values should be decreased and the number of points entered in the algorithm increased.

If in spite of all above, incorrect patterns are found, the initial value of the sample-detector distance should be slightly changed.

# CHAPTER 4

# PROGRAMMING AND USE OF THE SUPPLIED SOFTWARE

## 4.1 Code development

The simulated patterns shown in this book have been generated by the algorithms discussed in chapter 3. It is worth mentioning that the ever existing truncation error can be crucial in the present case. This is due to the type of operations employed in the algorithms and the very different order of magnitude of values encountered. The simplest way to avoid this is to employ a long data type length. However, this will increase the computing time and the memory requirement. To overcome this problem and to make the program more reliable for PCs, the data type length has been varied depending on the specific variable, the operation nature and the stage of the algorithms.

A graphic user interface has been developed based on an interactive system of guided menus. The input data necessary at each stage is requested on the screen. This makes the software extremely user friendly and can be used without any preliminary training.

The spot size considered in the simulation program has a diameter of 0.5 mm. Therefore, a precision of 12' can be achieved in the orientation if a fine collimation, a suitable detector and the typical sample-to-detector distance of 3cm are employed to carry out the experimental patterns.

For the implementation of the algorithm for the estimation of intensity, the source emission considered is the unfiltered and unpolarized from a Mo cathode tube operating at 20kV and the detector characteristics of a AgBr film.

## 4.2 Hardware specifications and installation procedure

To execute the program, a PC with a MS-DOS operating system, a VGA screen adaptator, a 3½ inch floppy drive and 465Kb of free conventional memory are required.

By typing `wslaue` in the MS-DOS command line of the 3½ inch supplied diskette the program is started. An example of a simulation and indexing work session is demonstrated in section 4.3.

It is strongly recommended to install the program in a hard disk partition (a minimum of 1.5 Mb free memory is required) using the installation program in the diskette. For installation, type `install`, from the 3½ inch drive with the floppy supplied. Then, enter the destination drive. The complete software will be installed in a new directory called `wsl`, which is created in the selected drive.

To execute the program, change to the `wsl` directory (`cd wsl`), and type `wslaue`. If any `Out of memory` message pops up, please refer to your operating system manuals in order to release conventional memory to a minimum of 465Kb.

The program can be installed and executed by Windows 3.1x, Windows 95 or Windows 98 by going into the Main group, opening a MS-DOS session and just following the steps mentioned above.

To get a 1:1 hard copy (9.25cm x 12.25cm) of the simulated patterns a PCL-4 or superior compatible printer, with a minimum of 1Mb of memory, must be plugged in the LPT1.

Usage of Pentium processor gives better performance in terms of computing time.


### 4.3 User manual

The use of the program is shown by a complete simulation and indexing session with intensity estimations for a $LiNbO_3$ sample. All the possible screens and menus will appear in the figures of this section.

The four stages of the session are:

4.3.1 Input of the geometrical data.
4.3.2 Selection of the job.
4.3.3 Simulation.
4.3.4 Indexing.

### 4.3.1 Input of the geometrical data

Once the program has been started by typing `wslaue` in the command line, the screen shown in figure 4.1 pops up.

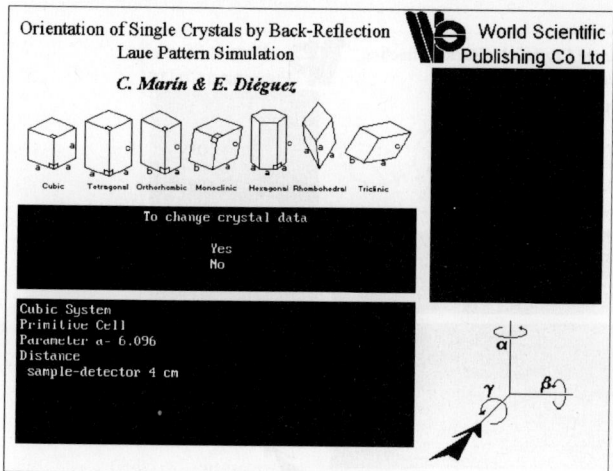

**Figure 4.1**

It presents the data of the experiment carried out in the last session. Now, either a session can be started with the same data by pressing N, or a new set of data can be entered by pressing Y. If N is pressed, the screen shown in figure 4.6 will pop up.

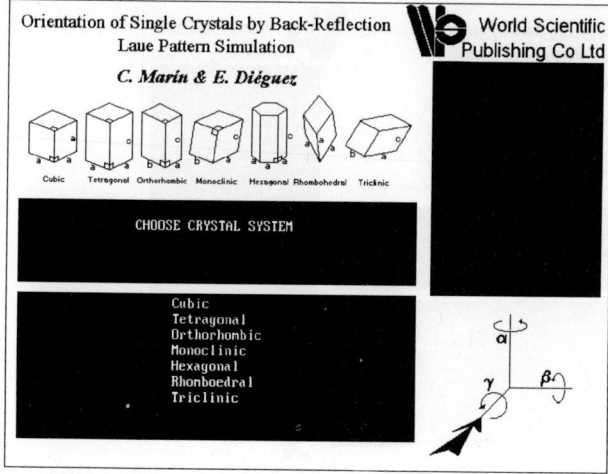

**Figure 4.2**

On pressing Y, the program will ask for the crystal system (figure 4.2), here we choose rhombohedral (R) for $LiNbO_3$.

For the rhombohedral system one can use the hexagonal by pressing Y in figure 4.3.

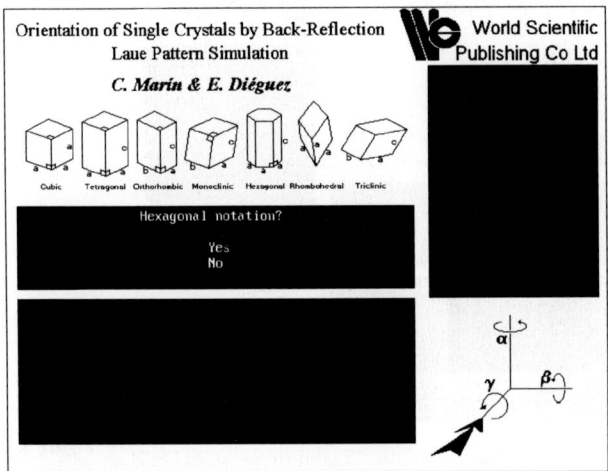

Figure 4.3

The lattice parameters and angles of the unit cell are typed in figure 4.4. The notation of lattice parameters sketched on the top of the screen should be followed while entering the data.

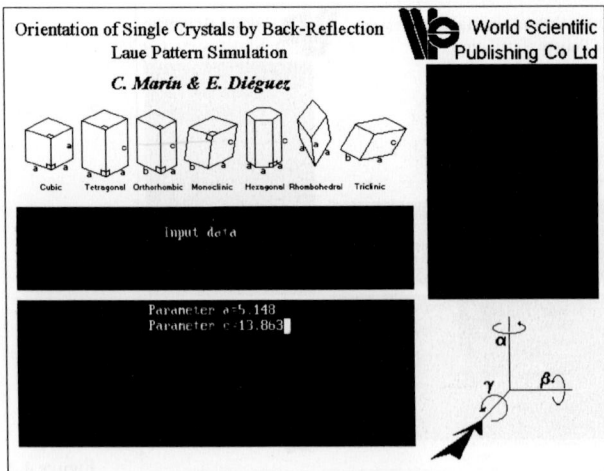

Figure 4.4

Finally, the distance (in cm) between sample and detector is entered in the screen of figure 4.5 (see section 2.1).

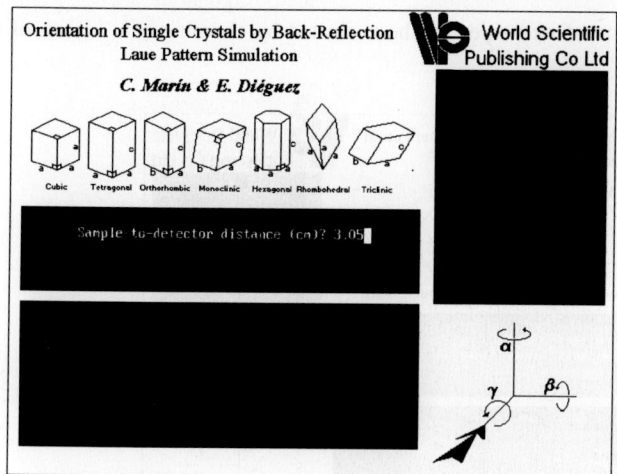

Figure 4.5

In order to confirm the input data, the screen of figure 4.1 will once again appear. To go to the next stage we press N.

## 4.3.2 Selection of the job

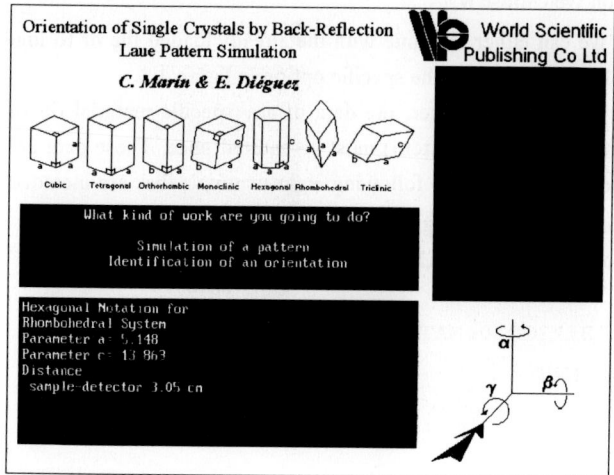

Figure 4.6

Once the geometrical data of the experiment are entered, the program asks for the kind of work to be carried out; either simulation or identification of a Laue-gram (figure 4.6).

The estimation of intensities can be evaluated by choosing W in screen of figure 4.7.

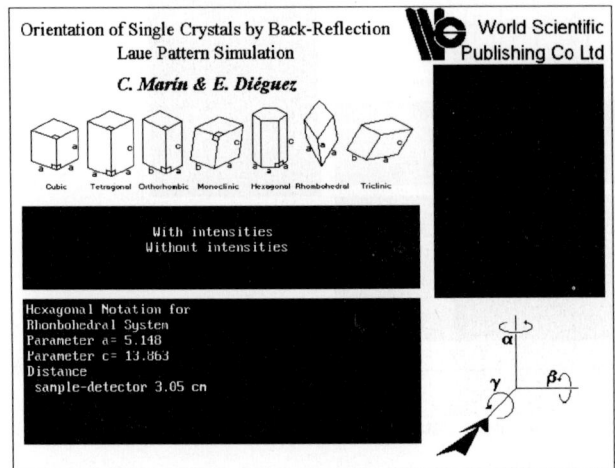

Figure 4.7

If I is pressed in figure 4.7, the screen shown in figure 4.15 or 4.23 will appear respectively depending on whether simulation or indexing was chosen in figure 4.6.

For this session we choose W.

At this stage we can either continue with the already loaded file or to load or to create a new material file by choosing the specific option in figure 4.8.

As discussed in previous chapter, the data of any specific material (in order to estimate the spots intensities of the pattern) must be entered once. These data are saved in a file with the extension .MAT. The following eight materials files are included in the supplied floppy diskette, corresponding to the materials for which the simulation results are shown in this book (see appendix A2):

$Bi_{12}GeO_{20}$ .- BI12GO20.MAT

GaSb .- GASB.MAT

$KH_2PO_4$ .- KH2PO4.MAT

$KTiOPO_4$ .- KTIOPO4.MAT

$(CH_2NH_2COOH)_3H_2SO_4$.- TGS.MAT

ZnO.-ZNO.MAT

$LiNbO_3$.- LINBO3.MAT

$Na_2W_4O_{13}$.- NA2W4O13.MAT

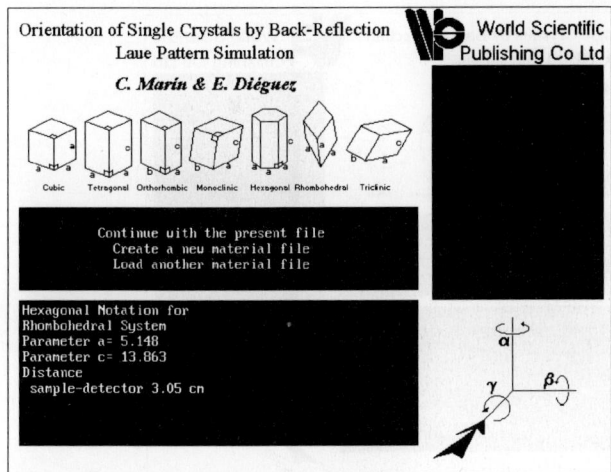

**Figure 4.8**

The example here shows how it has been done for LINBO3.MAT. Hence, if test is carried out by choosing R in figure 4.8, it is advisable to enter a different file name. Same file name will overwrite the present file.

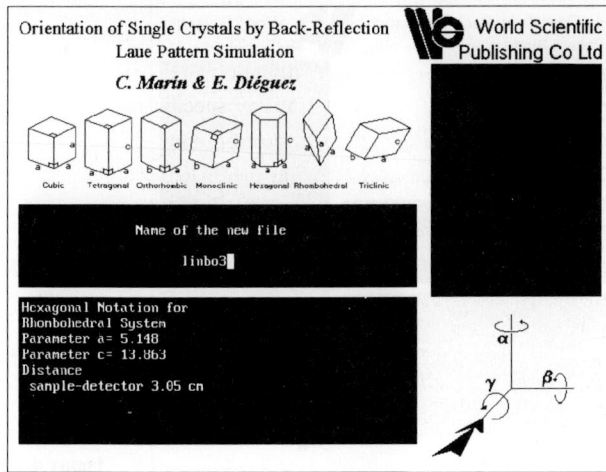

**Figure 4.9**

To create a file, a name will be asked, which will be saved with the extension .MAT (do not write the extension, it is automatically added, figure 4.9).

The Debye temperature will be asked (see section 2.3.4) as in figure 4.10. The value of Debye temperature for $LiNbO_3$ can be found in the appendix A2.7.

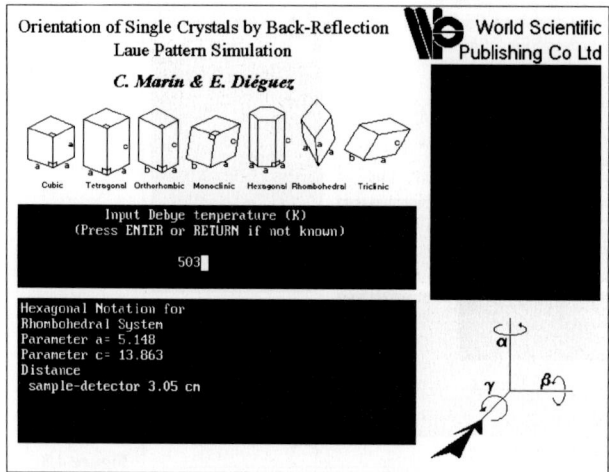

**Figure 4.10**

The total number of atoms per unit cell and the different kinds of atoms forming the structure are typed in figure 4.11 (see appendix A2.7).

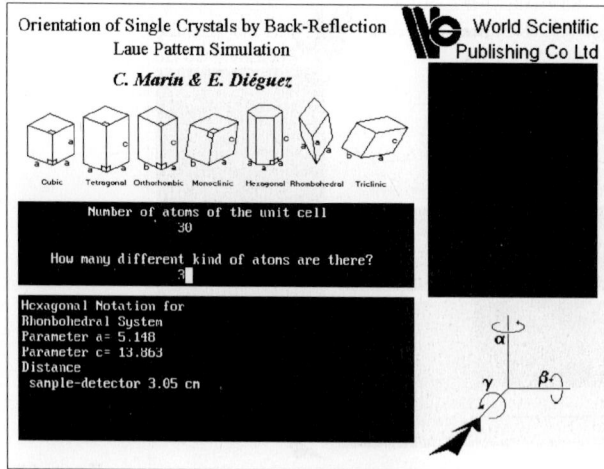

**Figure 4.11**

The atomic number and the number of each kind of atom per unit cell are specified in figure 4.12 (see appendix A2.7).

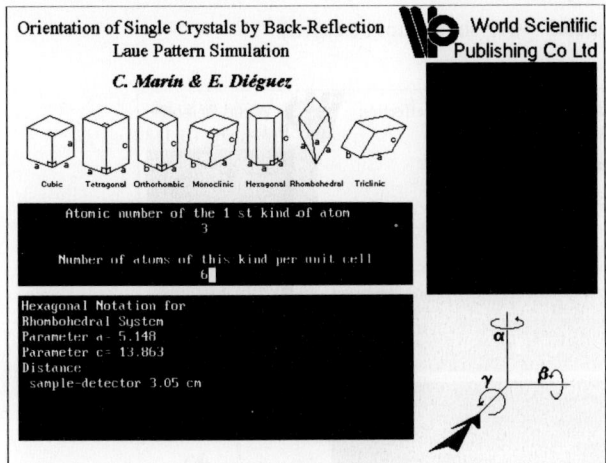

**Figure 4.12**

Then, the relative coordinates x, y and z of each of the atoms must be sequentially typed as requested by the program as shown in figure 4.13 (see appendix A2.7).

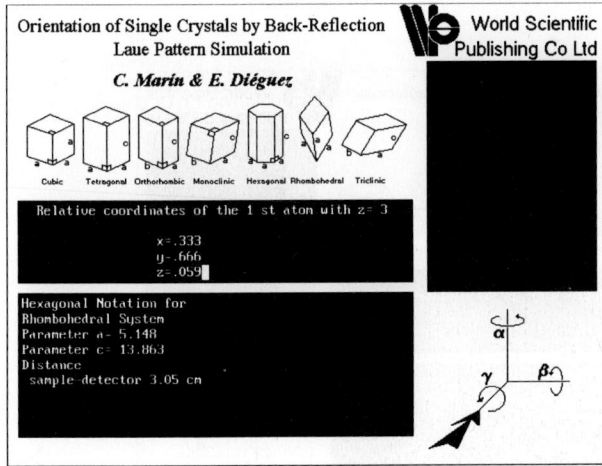

**Figure 4.13**

Finally, after an interpolation process set by the program itself, the file will be saved and the screen of figure 4.8 will appear again. Now we are going to load the LINBO3.MAT file by pressing L, we must type LINBO3 in the screen shown in the figure 4.14 (the extension .MAT should not be typed).

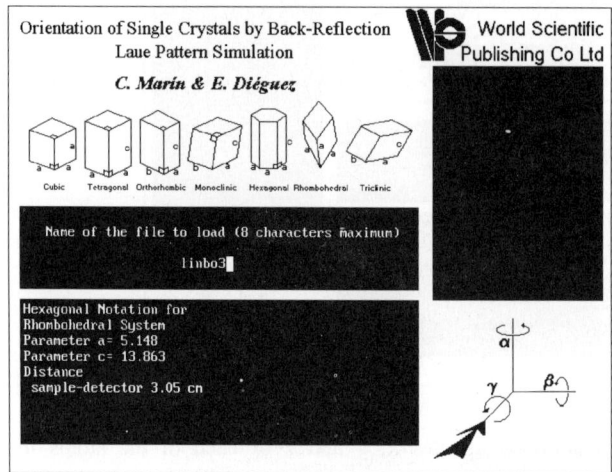

**Figure 4.14**

Once the file is loaded, the job selected in the screen of the figure 4.6 will start.

### 4.3.3 Simulation

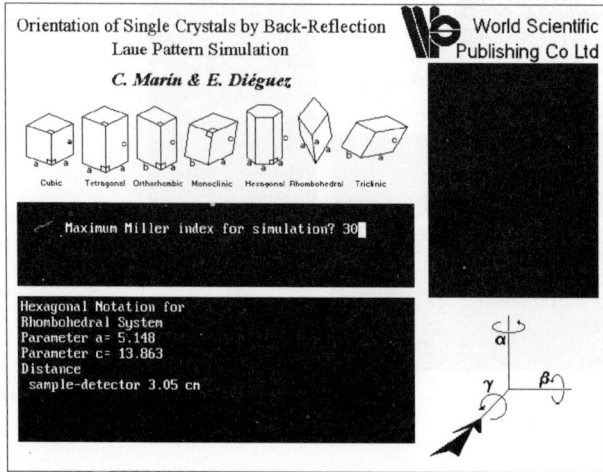

**Figure 4.15**

The maximum Miller index to be considered during the simulation will appear in figure 4.15.

In this step, we must proceed warily, because if a too high value is entered, and the spatial resolution between two spots is lower than the one in the program (0.5 millimetres), two spots will superimpose in the same position. If it happens, the program will consider the last spot and will overwrite the previous plane designation given to that position, resulting in an incorrect pattern.

In the next step, the orientation to simulate (figure 4.16) and the rotations α, β and γ (in this order) to be operated (figure 4.17) will be asked. In entering rotation angles, the rotation directions sketched in the screen must be followed.

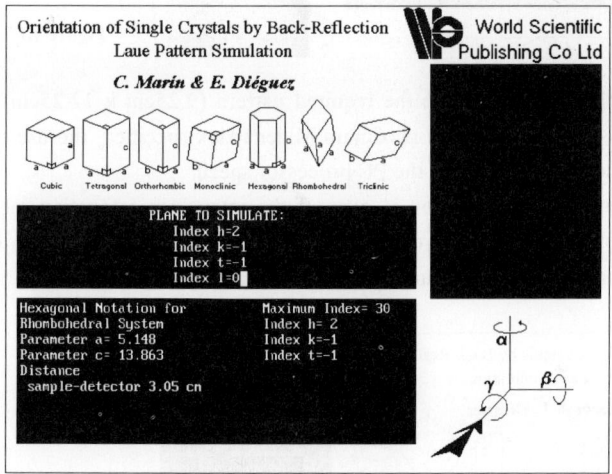

**Figure 4.16**

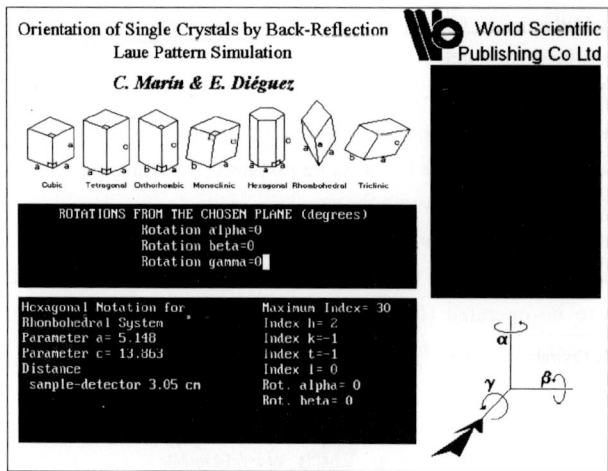

**Figure 4.17**

Now, the program will simulate the required pattern (9.25cm x 12.25cm). The time taken by this task depends on the maximum Miller index specified, the use of the estimation of intensities subroutine, and the PC processor speed.

When the geometrical simulation is over, if the intensities estimation job was chosen in the menu of figure 4.6, the program requires the detection an saturation levels (see section 3.2.2.a) in the screen of figure 4.18.

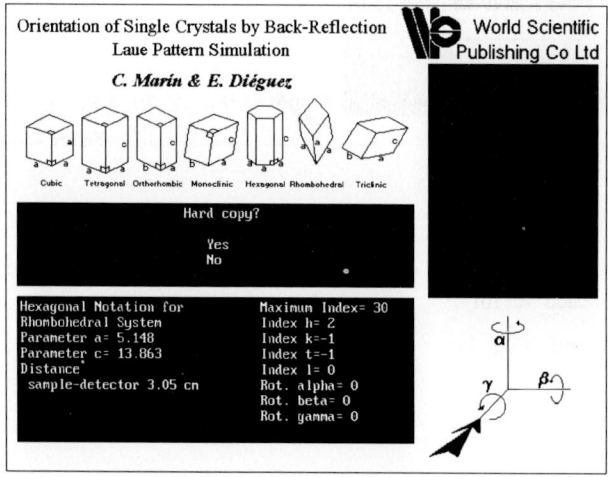

**Figure 4.18**

The Laue-gram will be shown in a eleven levels grey-scale, not showing the spots with intensity estimation below the detection level and printing the same highest intensity above the saturation level. These levels can be changed continuously (by typing F) to closely match the experimental pattern.

Then, the option to get a hard copy of the pattern (figure 4.19) in a one to one scale and with six intensity levels (see section 3.2.2.a) is provided.

If the specific printer mentioned in section 4.2 is not connected, press N.

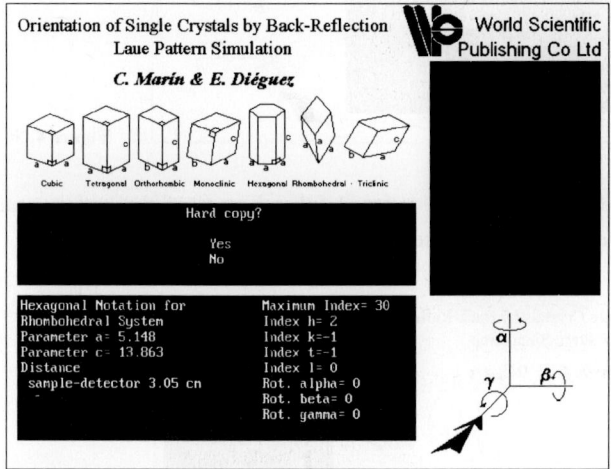

Figure 4.19

At this point (figure 4.20), the label of each spot in the simulated Laue-gram can be printed. In a hard copy the indices of the plane assigned to each spot. In this pattern, the three indices h, k and l are written, even if the hexagonal notation is being employed (see section 1.3.2). One digit is shown for each index, being added a point above it: either on the left side if it must be added 10, or in the centre if it must be added 20 or in the right if it must be added 30; for example 13 index is written ˙3, $\overline{22}$ index is written $\overline{2}$, 7 index is written 7, and so on.

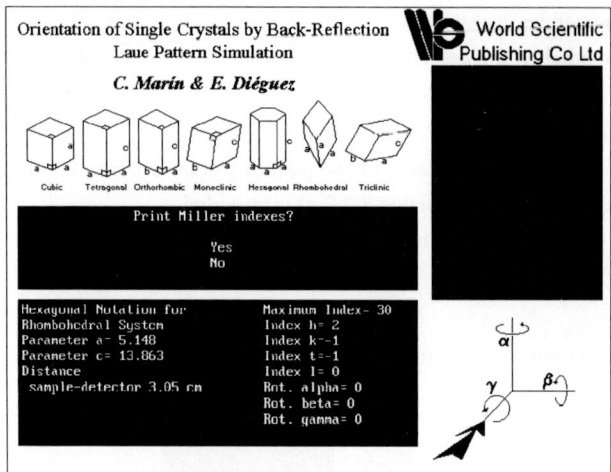

Figure 4.20

Finally (figure 4.21), the plane assigned to each spot of the pattern, and the intensity level can be obtained on the screen.

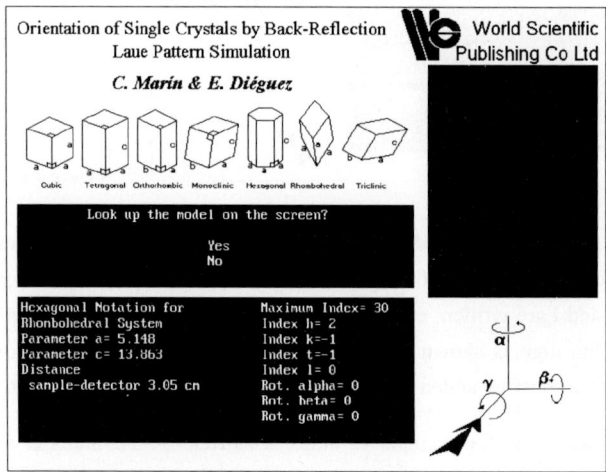

Figure 4.21

In the screen of figure 4.22, by superimposing the cursor on any of the spots of the pattern the information will be displayed. The cursor precision can be increased or decreased as indicated on the screen.

**Figure 4.22**

### 4.3.4 Indexing

If the identification option was chosen in the menu of figure 4.6, the screen of figure 4.23 will appear asking about the maximum Miller index to select the range of planes to be tested for the two fundamental spots to be entered in the algorithm (see section 3.3.3).

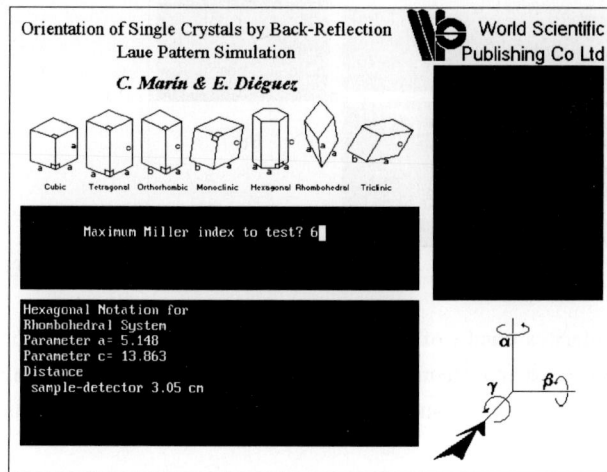

**Figure 4.23**

Then, the values for the angular (figure 4.24) and the index (figure 4.25) tolerance are entered (see section 3.3.3).

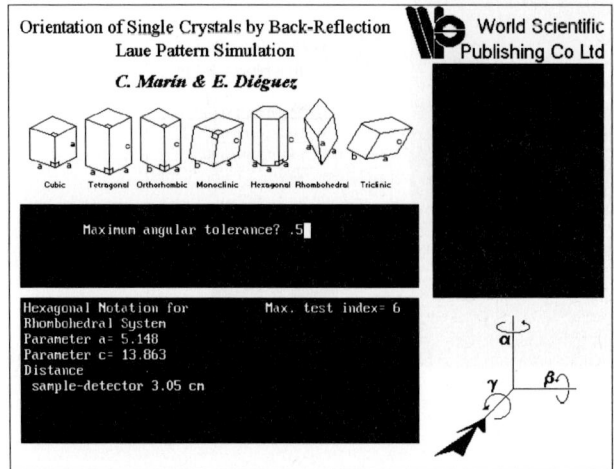

**Figure 4.24**

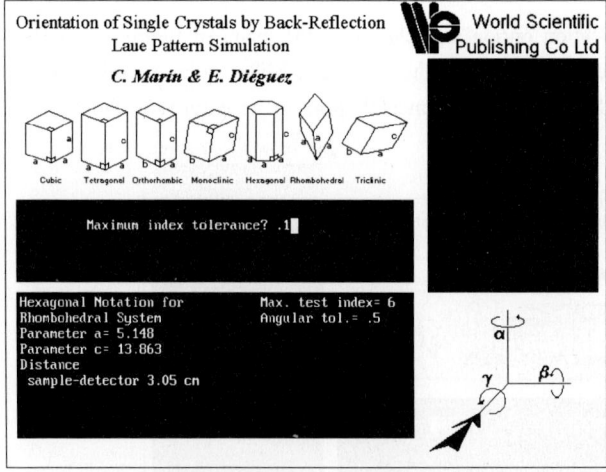

**Figure 4.25**

Next, the coordinates x and y of the experimental Laue-gram selected to perform the indexing algorithm are entered (figure 4.26), see section 3.3.2. By entering no values for x and y, the program ends the input stage, while the indexing algorithm starts to be executed.

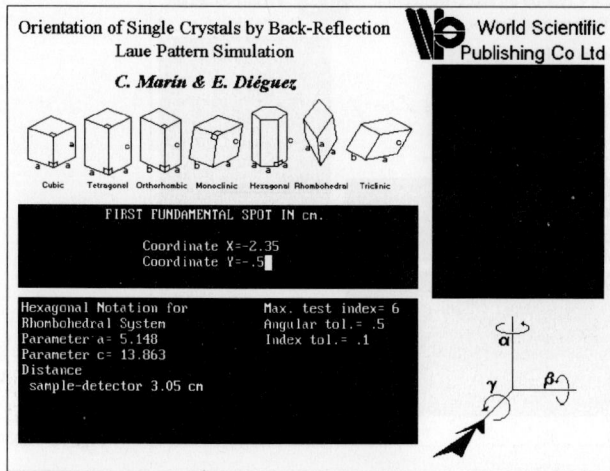

**Figure 4.26**

Finally, the number of solutions, and the option to check the planes assigned by each solution to the experimental spots is given (figures 4.27 and 4.28). For each solution, the simulated Laue pattern can be visualised by pressing S in menu of figure 4.28 and typing S in figure 4.29.

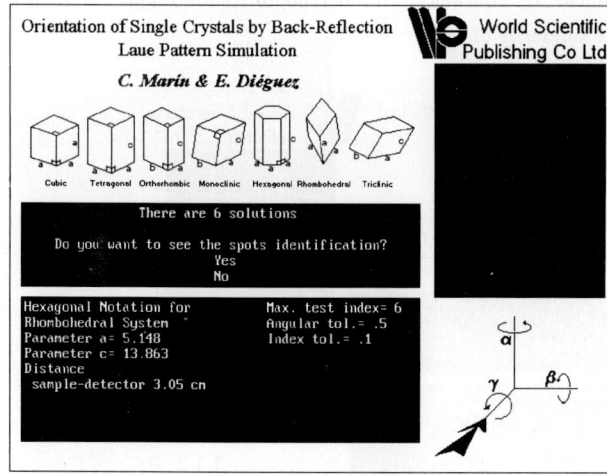

**Figure 4.27**

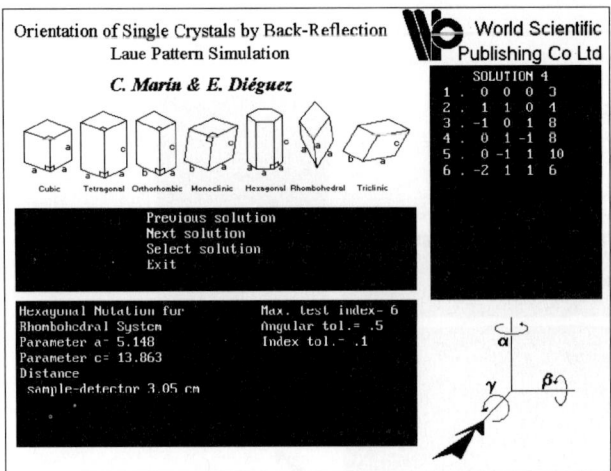

**Figure 4.28**

Once the correct identification is done by comparison between the experimental and the simulated Laue-grams, the rotation angles to be set to locate the sample in any desired orientation can be obtained by pressing G in the screen of figure 4.29 and entering the orientation in the screen of figure 4.30. Four options are given according to different combinations of rotation axes (figure 4.31). We can chose the most suitable one for the specific experiment, by considering the maximum rotation angle of each axis of the goniometer and the solidarity of its movements.

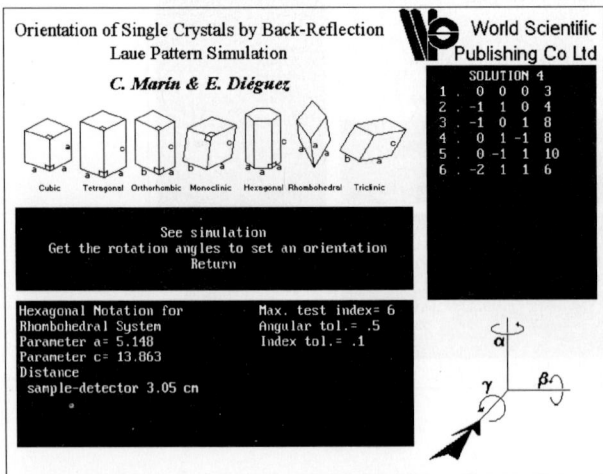

**Figure 4.29**

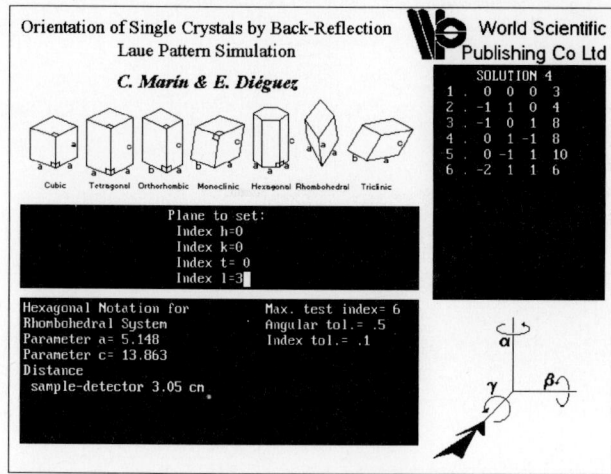

**Figure 4.30**

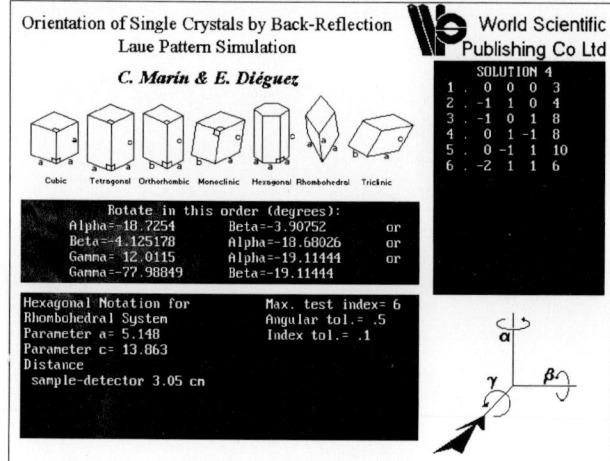

**Figure 4.31**

By pressing E in the screen of figure 4.29, the indexing session ends. It should be press Esc in any of the screens to finish the program execution.

CHAPTER 5

## COMPARISON AND DISCUSSION BETWEEN SIMULATED AND EXPERIMENTAL PATTERNS OF SAMPLES OF THE SEVEN CRYSTAL LATTICE

In this chapter we are going to show examples of results of the procedures discussed in chapter 4. Seven crystals of each crystal lattice are going to be employed. Each section begins with an experimental non-oriented Laue-gram of the material. The selected data entered in the program will be shown. Once the identification of the pattern is done, three orientations of each crystal will be chosen and both patterns, the experimental and the simulated, will be compared.

### 5.1 Cubic: GaSb

Figure 5.1a shows an experimental Laue-gram of an unknown orientation of a GaSb single crystal (see appendix A2.2). To proceed with the indexing of the Laue-gram, points 1 to 3 are selected, considering spots 1 and 2 as fundamental, according with the rules discussed in section 3.3.1. Then, by introducing into the algorithm the indexing data shown in table 5.1, the 24 solutions shown in table 5.2 are suggested.

The simulated Laue-gram shown in figure 5.1b corresponds to all indexing solutions presented in table 5.2 for the symmetry involved.

**Table 5.1:** *Inputs for the indexing of the Laue-gram in figure 5.1a.*

| | x | y |
|---|---|---|
| Maximum Miller index | 1 | |
| Sample-to-detector distance (cm) | 3.1 | |
| Maximum angular tolerance (°) | 1 | |
| Maximum index tolerance | 0.1 | |
| | x | y |
| Coordinates of fundamental spots (cm) | | |
| Spot 1 | 0.600 (25) | -2.550 (25) |
| Spot 2 | -3.700 (25) | 0.000 (25) |
| Coordinates of the other spots (cm) | | |
| Spot 3 | 1.700 (25) | -0.350 (25) |

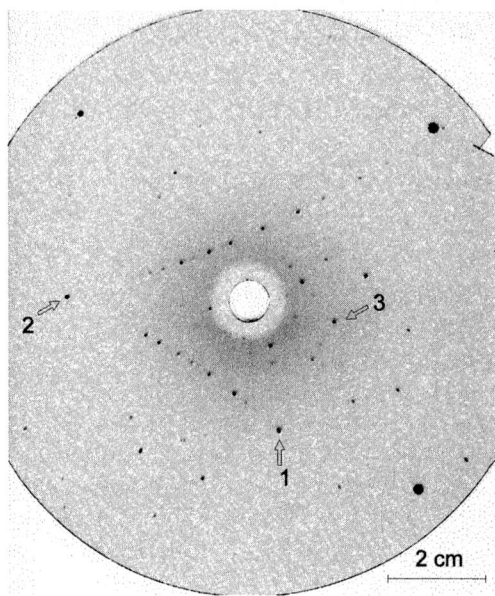

2 cm

**Figure 5.1a:** *Experimental back-reflection Laue-gram for GaSb: Mo tube V=20kV, I=40mA, exposure time 180 min., Kodak AX film, sample-to-detector distance 3.1 cm.*

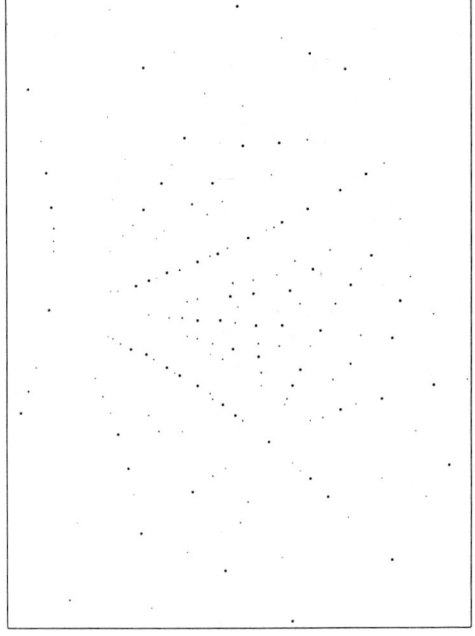

**Figure 5.1b:** *Simulated back-reflection Laue-gram for GaSb. Solution 23 in table 5.2; maximum Miller index 12, detection level 0.1%, saturation level 5%, sample-to-detector distance 3.1 cm.*

**Table 5.2:** *Indexing solutions for the Laue-gram in figure 5.1a with the inputs of table 5.1*

|   | Solution 1 | Solution 2 | Solution 3 | Solution 4 |
|---|---|---|---|---|
| 1 | $\bar{1}\,\bar{1}\,0$ | $\bar{1}\,\bar{1}\,0$ | $\bar{1}\,0\,\bar{1}$ | $\bar{1}\,0\,\bar{1}$ |
| 2 | $\bar{1}\,1\,\bar{1}$ | $\bar{1}\,1\,\bar{1}$ | $\bar{1}\,1\,\bar{1}$ | $\bar{1}\,1\,\bar{1}$ |
| 3 | $\bar{1}\,\bar{2}\,0$ | $\bar{2}\,\bar{1}\,0$ | $\bar{2}\,0\,\bar{1}$ | $\bar{1}\,0\,\bar{2}$ |

|   | Solution 5 | Solution 6 | Solution 7 | Solution 8 |
|---|---|---|---|---|
| 1 | $\bar{1}\,0\,1$ | $\bar{1}\,0\,1$ | $\bar{1}\,1\,0$ | $\bar{1}\,1\,0$ |
| 2 | $\bar{1}\,1\,1$ | $\bar{1}\,1\,1$ | $\bar{1}\,1\,1$ | $\bar{1}\,1\,1$ |
| 3 | $\bar{1}\,0\,2$ | $\bar{2}\,0\,1$ | $\bar{2}\,1\,0$ | $\bar{1}\,2\,0$ |

|   | Solution 9 | Solution 10 | Solution 11 | Solution 12 |
|---|---|---|---|---|
| 1 | $0\,\bar{1}\,\bar{1}$ | $0\,\bar{1}\,\bar{1}$ | $0\,\bar{1}\,1$ | $0\,\bar{1}\,1$ |
| 2 | $\bar{1}\,\bar{1}\,\bar{1}$ | $1\,\bar{1}\,\bar{1}$ | $\bar{1}\,\bar{1}\,1$ | $1\,\bar{1}\,1$ |
| 3 | $0\,\bar{1}\,\bar{2}$ | $0\,\bar{2}\,\bar{1}$ | $0\,\bar{2}\,1$ | $0\,\bar{1}\,2$ |

|   | Solution 13 | Solution 14 | Solution 15 | Solution 16 |
|---|---|---|---|---|
| 1 | $0\,1\,\bar{1}$ | $0\,1\,\bar{1}$ | $0\,1\,1$ | $0\,1\,1$ |
| 2 | $\bar{1}\,1\,\bar{1}$ | $1\,1\,\bar{1}$ | $\bar{1}\,1\,1$ | $1\,1\,1$ |
| 3 | $0\,2\,\bar{1}$ | $0\,1\,\bar{2}$ | $0\,1\,2$ | $0\,2\,1$ |

|   | Solution 17 | Solution 18 | Solution 19 | Solution 20 |
|---|---|---|---|---|
| 1 | $1\,\bar{1}\,0$ | $1\,\bar{1}\,0$ | $1\,0\,\bar{1}$ | $1\,0\,\bar{1}$ |
| 2 | $1\,\bar{1}\,\bar{1}$ | $1\,\bar{1}\,1$ | $1\,\bar{1}\,\bar{1}$ | $1\,1\,\bar{1}$ |
| 3 | $2\,\bar{1}\,0$ | $1\,\bar{2}\,0$ | $1\,0\,\bar{2}$ | $2\,0\,\bar{1}$ |

|   | Solution 21 | Solution 22 | Solution 23 | Solution 24 |
|---|---|---|---|---|
| 1 | $1\,0\,1$ | $1\,0\,1$ | $1\,1\,0$ | $1\,1\,0$ |
| 2 | $1\,\bar{1}\,1$ | $1\,1\,1$ | $1\,1\,\bar{1}$ | $1\,1\,1$ |
| 3 | $2\,0\,1$ | $1\,0\,2$ | $1\,2\,0$ | $2\,1\,0$ |

Four combinations of angles to get the orientations (0 0 1), (0 1 1) and (1 1̄ 1)
are calculated from the solution 23 shown in table 5.2 (table 5.3). The experimental and
simulated Laue-grams shown in figures 5.2, 5.3 and 5.4 are obtained by applying the
rotation combination α and β. As expected, they correspond to the orientations (0 0 1),
(0 1 1) and (1 1̄ 1), respectively.

**Table 5.3:** *Rotation angles (four combinations) calculated to obtain the orientations (0 0 1), (0 1 1) and (1 1̄ 1) from the solution 23 in table 5.2.*

|  | (0 0 1) | (0 1 1) | (1 1̄ 1) |
|---|---|---|---|
| α | 105.7° | 64.8° | 169.4° |
| β | -28.1° | -5.3° | -69.8° |
| β | 63.1° | -12.3° | 70.1° |
| α | 121.9° | 64.3° | 176.3° |
| γ | -29.0° | -5.8° | -86.1° |
| α | -256.2° | 64.9° | -250.1° |
| γ | 61.0° | 84.2° | 3.9° |
| β | 103.8° | -64.9° | 109.8° |

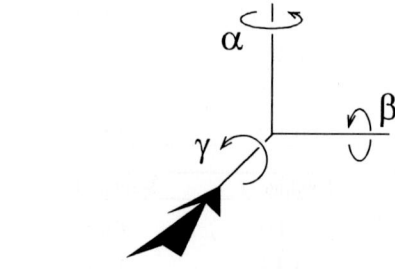

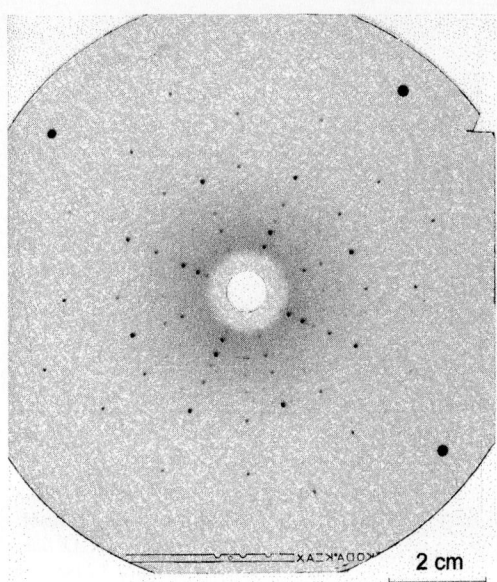

**Figure 5.2a:** *Experimental back-reflection Laue-gram for GaSb: plane (0 0 1), Mo tube V=20kV, I=40mA, exposure time 180 min., Kodak AX film, sample-to-detector distance 3.1 cm.*

2 cm

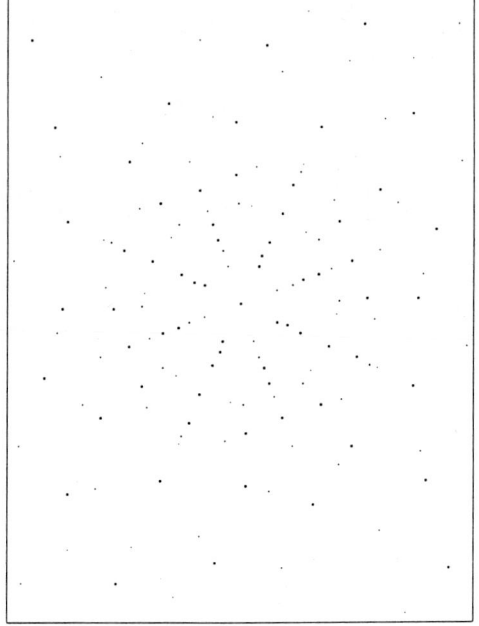

**Figure 5.2b:** *Simulated back-reflection Laue-gram for GaSb, plane (0 0 1), maximum Miller index 12, detection level 0.01%, saturation level 4%, sample-to-detector distance 3.1 cm.*

2 cm

**Figure 5.3a:** *Experimental back-reflection Laue-gram for GaSb: plane (0 1 1), Mo tube V=20kV, I=40mA, exposure time 180 min., Kodak AX film, sample-to-detector distance 3.1 cm.*

**Figure 5.3b:** *Simulated back-reflection Laue-gram for GaSb, plane (0 1 1), maximum Miller index 15, detection level 0.01%, saturation level 8.1%, sample-to-detector distance 3.1 cm.*

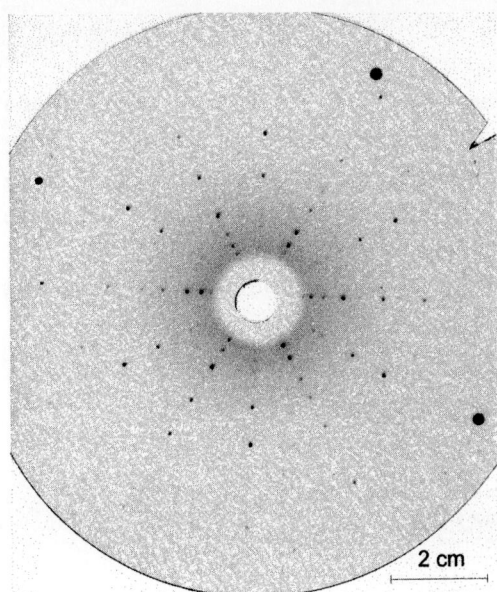

2 cm

**Figure 5.4a:** *Experimental back-reflection Laue-gram for GaSb: plane (1 1̄ 1), Mo tube V=20kV, I=40mA, exposure time 180 min., Kodak AX film, sample-to-detector distance 3.1 cm.*

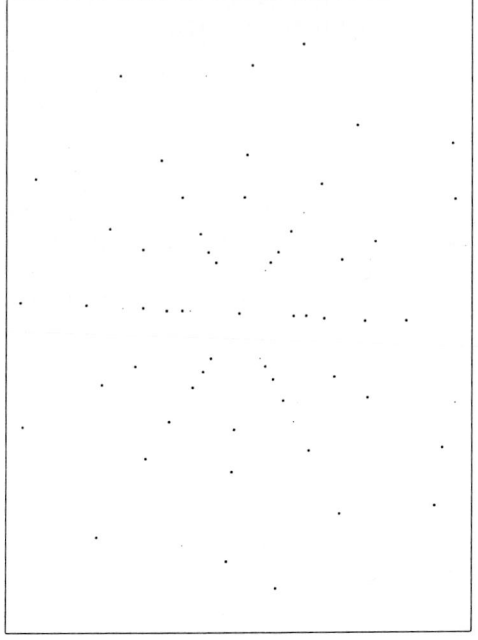

**Figure 5.4b:** *Simulated back-reflection Laue-gram for GaSb, plane (1 1̄ 1), maximum Miller index 14, detection level 1%, saturation level 20%, sample-to-detector distance 3.1 cm.*

## 5.2 Tetragonal: $KH_2PO_4$

Figure 5.5a shows an experimental Laue-gram of an unknown orientation of a $KH_2PO_4$ single crystal (see appendix A2.3). To proceed with the indexing of the Laue-gram, points 1 to 4 are selected, considering spots 1 and 2 as fundamental, according with the rules discussed in section 3.3.1. Then, by introducing into the algorithm the indexing data shown in table 5.4, the 8 solutions shown in table 5.5 are suggested.

The simulated Laue-gram shown in figure 5.5b corresponds to all indexing solutions presented in table 5.5 for the symmetry involved.

**Table 5.4:** *Inputs for the indexing of the Laue-gram in figure 5.5a.*

| | | |
|---|---|---|
| Maximum Miller index | 1 | |
| Sample-to-detector distance (cm) | 3.2 | |
| Maximum angular tolerance (°) | 1 | |
| Maximum index tolerance | 0.1 | |
| | x | y |
| Coordinates of fundamental spots (cm) | | |
| Spot 1 | 0.650 (25) | -2.850 (25) |
| Spot 2 | 0.300 (25) | 2.100 (25) |
| Coordinates of the other spots (cm) | | |
| Spot 3 | -1.750 (25) | 0.500 (25) |
| Spot 4 | 0.250 (25) | 0.700 (25) |

**Table 5.5:** *Indexing solutions for the Laue-gram in figure 5.5a with the inputs of table 5.4.*

| | Solution 1 | Solution 2 | Solution 3 | Solution 4 |
|---|---|---|---|---|
| 1 | $\bar{1}\bar{1}0$ | $\bar{1}\bar{1}0$ | $\bar{1}10$ | $\bar{1}10$ |
| 2 | $\bar{1}\bar{1}\bar{1}$ | $\bar{1}\bar{1}1$ | $\bar{1}1\bar{1}$ | $\bar{1}11$ |
| 3 | $\bar{2}\bar{1}\bar{1}$ | $\bar{1}2\bar{1}$ | $\bar{1}2\bar{1}$ | $\bar{2}11$ |
| 4 | $\bar{3}\bar{3}\bar{2}$ | $\bar{3}3\bar{2}$ | $\bar{3}3\bar{2}$ | $\bar{3}32$ |

| | Solution 5 | Solution 6 | Solution 7 | Solution 8 |
|---|---|---|---|---|
| 1 | $1\bar{1}0$ | $1\bar{1}0$ | $110$ | $110$ |
| 2 | $1\bar{1}\bar{1}$ | $1\bar{1}1$ | $11\bar{1}$ | $111$ |
| 3 | $1\bar{2}\bar{1}$ | $2\bar{1}1$ | $21\bar{1}$ | $121$ |
| 4 | $3\bar{3}\bar{2}$ | $3\bar{3}2$ | $33\bar{2}$ | $332$ |

**Figure 5.5a:** *Experimental back-reflection Laue-gram for KH$_2$PO$_4$: Mo tube V=20kV, I=40mA, exposure time 180 min., Kodak AX film, sample-to-detector distance 3.2 cm.*

2 cm

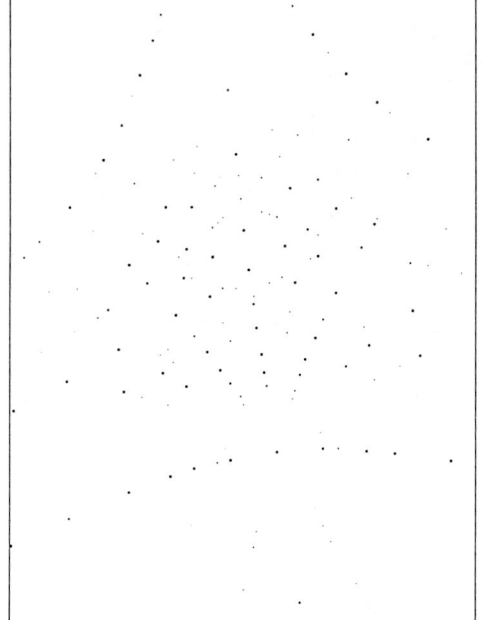

**Figure 5.5b:** *Simulated back-reflection Laue-gram for KH$_2$PO$_4$. Solution 8 in table 5.5; maximum Miller index 12, detection level 1%, saturation level 17%, sample-to-detector distance 3.2 cm.*

Four combinations of angles to get the orientations (1 0 1), (1 0 0) and (1 1 1) are calculated from the solution 8 shown in table 5.5 (table 5.6). The experimental and simulated Laue-grams shown in figures 5.6, 5.7 and 5.8 are obtained by applying the rotation combination α and β. As expected, they correspond to the orientations (1 0 1), (1 0 0) and (1 1 1), respectively.

**Table 5.6:** *Rotation angles (four combinations) calculated to obtain the orientations (1 0 1), (1 0 0) and (1 1 1) from the solution 8 in table 5.5.*

|   | $(1\,0\,1)$ | $(1\,0\,0)$ | $(1\,1\,1)$ |
|---|---|---|---|
| α | 35.8° | 51.0° | 2.4° |
| β | 32.9° | -11.8° | 16.4° |
| β | 38.6° | -18.3° | 16.4° |
| α | 29.4° | 49.6° | 2.3° |
| γ | 47.8° | -15.0° | 81.7° |
| α | 47.1° | 52.0° | 16.6° |
| γ | -42.1° | 75.0° | -8.3° |
| β | 47.1° | -52.0° | 16.6° |

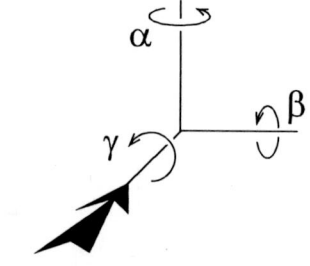

**Figure 5.6a:** *Experimental back-reflection Laue-gram for $KH_2PO_4$: plane (1 0 1), Mo tube V=20kV, I=40mA, exposure time 180 min., Kodak AX film, sample-to-detector distance 3.2 cm.*

2 cm

**Figure 5.6b:** *Simulated back-reflection Laue-gram for $KH_2PO_4$, plane (1 0 1), maximum Miller index 15, detection level 1%, saturation level 40%, sample-to-detector distance 3.2 cm.*

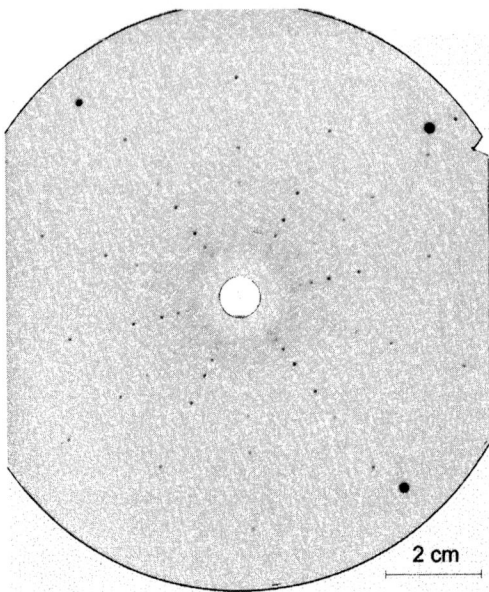

2 cm

**Figure 5.7a:** *Experimental back-reflection Laue-gram for $KH_2PO_4$: plane (1 0 0), Mo tube V=20kV, I=40mA, exposure time 180 min., Kodak AX film, sample-to-detector distance 3.05 cm.*

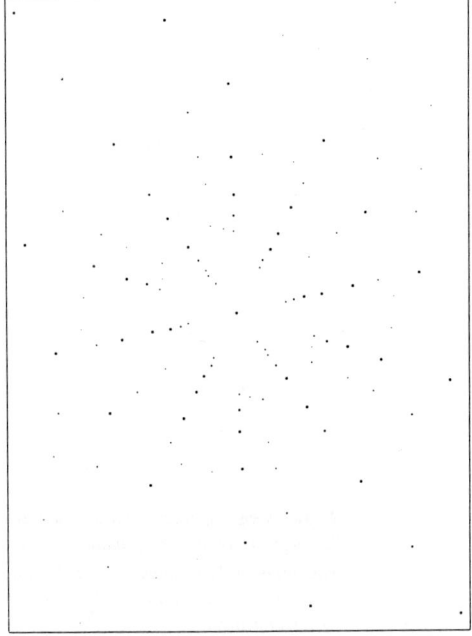

**Figure 5.7b:** *Simulated back-reflection Laue-gram for $KH_2PO_4$, plane (1 0 0), maximum Miller index 15, detection level 1%, saturation level 30%, sample-to-detector distance 3.05 cm.*

**Figure 5.8a:** *Experimental back-reflection Laue-gram for KH₂PO₄: plane (1 1 1), Mo tube V=20kV, I=40mA, exposure time 180 min., Kodak AX film, sample-to-detector distance 3.05 cm.*

2 cm

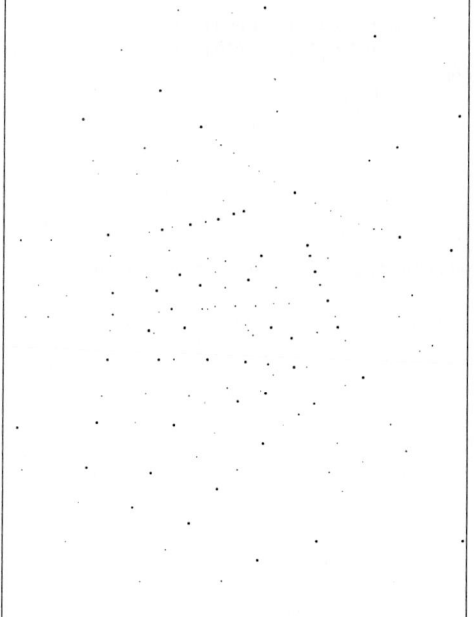

**Figure 5.8b:** *Simulated back-reflection Laue-gram for KH₂PO₄, plane (1 1 1), maximum Miller index 14, detection level 1%, saturation level 20%, sample-to-detector distance 3.05 cm.*

## 5.3 Orthorhombic: KTiOPO$_4$

Figure 5.9a shows an experimental Laue-gram of an unknown orientation of a KH$_2$PO$_4$ single crystal (see appendix A2.4). To proceed with the indexing of the Laue-gram, points 1 to 4 are selected, considering spots 1 and 2 as fundamental, according with the rules discussed in section 3.3.1. Then, by introducing into the algorithm the indexing data shown in table 5.7, the 8 solutions shown in table 5.8 are suggested.

The simulated Laue-gram shown in figure 5.9b corresponds to all indexing solutions presented in table 5.8 for the symmetry involved.

**Table 5.7:** *Inputs for the indexing of the Laue-gram in figure 5.9a.*

| | x | y |
|---|---|---|
| Maximum Miller index | 1 | |
| Sample-to-detector distance (cm) | 3.2 | |
| Maximum angular tolerance (°) | 1 | |
| Maximum index tolerance | 0.1 | |
| Coordinates of fundamental spots (cm) | | |
| Spot 1 | 0.350 (25) | 1.600 (25) |
| Spot 2 | -0.900 (25) | -0.950 (25) |
| Coordinates of the other spots (cm) | | |
| Spot 3 | 2.700 (25) | -4.650 (25) |
| Spot 4 | 2.000 (25) | 0.800 (25) |

**Table 5.8:** *Indexing solutions for the Laue-gram in figure 5.9a with the inputs of table 5.7.*

| | Solution 1 | Solution 2 | Solution 3 | Solution 4 |
|---|---|---|---|---|
| 1 | 0 $\bar{1}$ $\bar{1}$ | 0 $\bar{1}$ 1 | 0 1 $\bar{1}$ | 0 1 1 |
| 2 | $\bar{1}$ $\bar{1}$ $\bar{1}$ | $\bar{1}$ $\bar{1}$ 1 | 1 1 $\bar{1}$ | $\bar{1}$ 1 1 |
| 3 | $\bar{1}$ $\bar{1}$ 0 | 1 $\bar{1}$ 0 | 1 1 0 | $\bar{1}$ 1 0 |
| 4 | 0 $\bar{2}$ $\bar{1}$ | 0 $\bar{2}$ 1 | 0 2 $\bar{1}$ | 0 2 1 |

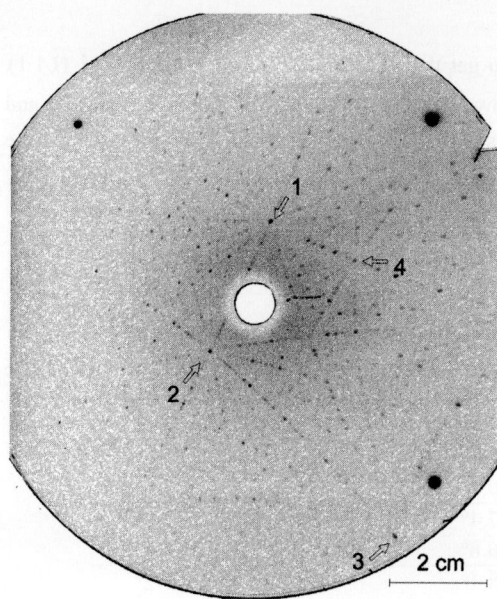

**Figure 5.9a:** *Experimental back-reflection Laue-gram for KTiOPO₄: Mo tube V=20kV, I=40mA, exposure time 210 min., Kodak AX film, sample-to-detector distance 3.2 cm.*

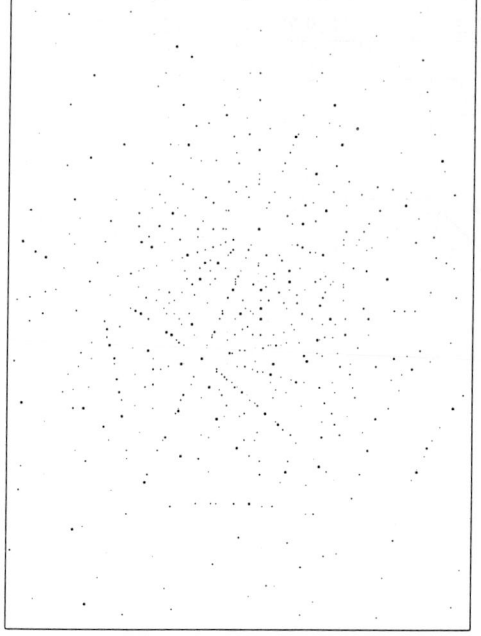

**Figure 5.9b:** *Simulated back-reflection Laue-gram for KTiOPO₄. Solution 4 in table 5.8; maximum Miller index 12, detection level 1%, saturation level 30%, sample-to-detector distance 3.2 cm.*

Four combinations of angles to get the orientations (1 1 0), (0 1 0) and ($\bar{1}$ 1 1) are calculated from the solution 2 shown in table 5.8 (table 5.9). The experimental and simulated Laue-grams shown in figures 5.10, 5.11 and 5.12 are obtained by applying the rotation combination $\alpha$ and $\beta$. As expected, they correspond to the orientations (1 1 0), (0 1 0) and ($\bar{1}$ 1 1), respectively.

**Table 5.9:** *Rotation angles (four combinations) calculated to obtain the orientations (1 1 0), (0 1 0) and ($\bar{1}$ 1 1) from the solution 2 in table 5.8.*

|  | ( 1 1 0 ) | ( 0 1 0 ) | ( $\bar{1}$ 1 1 ) |
|---|---|---|---|
| $\alpha$ | 224.2° | 210.2° | 225.6° |
| $\beta$ | -20.4° | 2.3° | 37.6° |
| $\beta$ | 27.4° | -2.6° | -47.8° |
| $\alpha$ | 220.8° | 210.2° | 214.4° |
| $\gamma$ | 28.1° | -4.5° | -47.2° |
| $\alpha$ | 227.8° | 210.3° | 236.3° |
| $\gamma$ | -61.9° | 85.5° | 42.8° |
| $\beta$ | -132.2° | 149.7° | 123.6° |

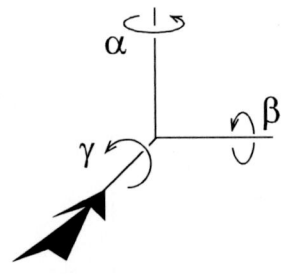

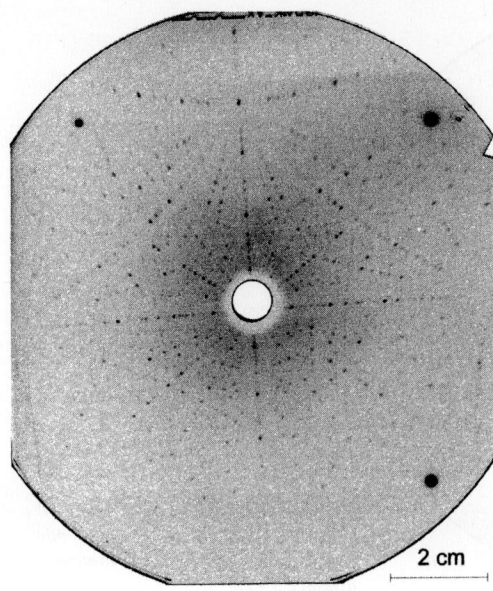

2 cm

**Figure 5.10a:** *Experimental back-reflection Laue-gram for KTiOPO₄: plane (1 1 0), Mo tube V=20kV, I=40mA, exposure time 210 min., Kodak AX film, sample-to-detector distance 3.1 cm.*

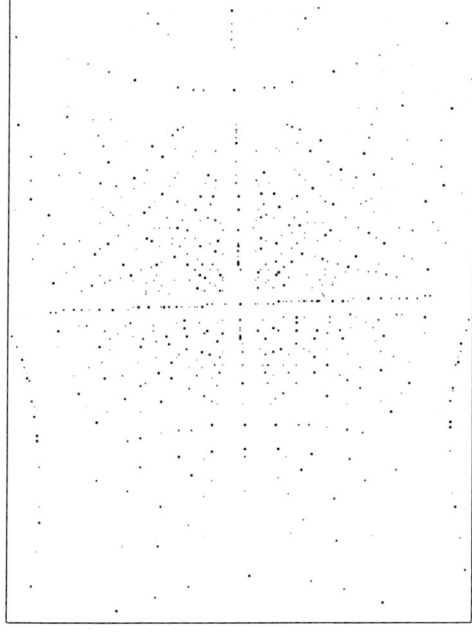

**Figure 5.10b:** *Simulated back-reflection Laue-gram for KTiOPO₄, plane (1 1 0), maximum Miller index XX, detection level 1%, saturation level 16%, sample-to-detector distance 3.1 cm.*

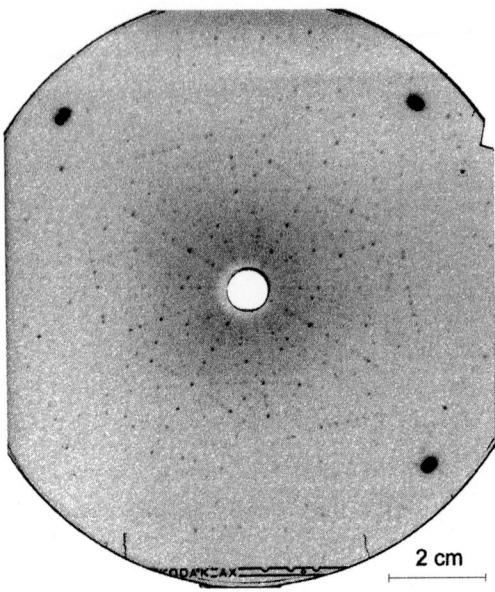

2 cm

**Figure 5.11a:** *Experimental back-reflection Laue-gram for KTiOPO₄: plane (0 1 0), Mo tube V=20kV, I=40mA, exposure time 210 min., Kodak AX film, sample-to-detector distance 3.4 cm.*

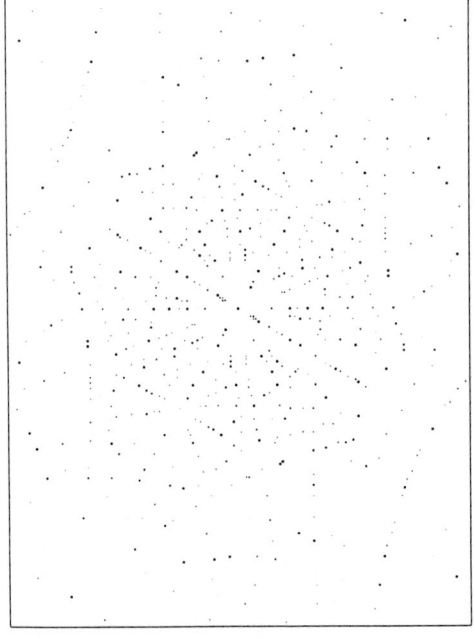

**Figure 5.11b:** *Simulated back-reflection Laue-gram for KTiOPO₄, plane (0 1 0), maximum Miller index 12, detection level 1%, saturation level 20%, sample-to-detector distance 3.4 cm.*

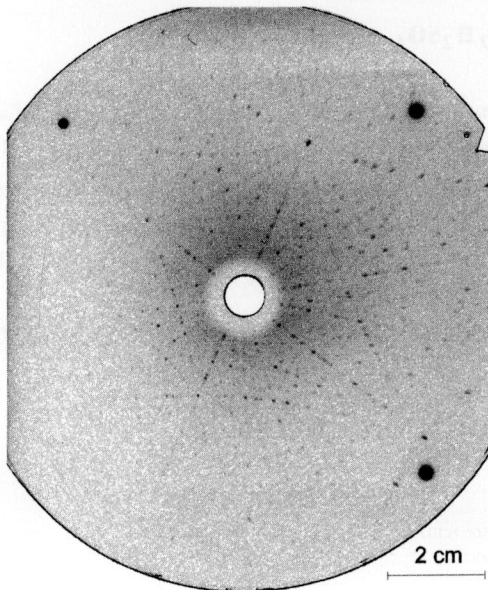

**Figure 5.12a:** *Experimental back-reflection Laue-gram for KTiOPO$_4$: plane $(\bar{1}\,1\,1)$, Mo tube V=20kV, I=40mA, exposure time 210 min., Kodak AX film, sample-to-detector distance 3.1 cm.*

2 cm

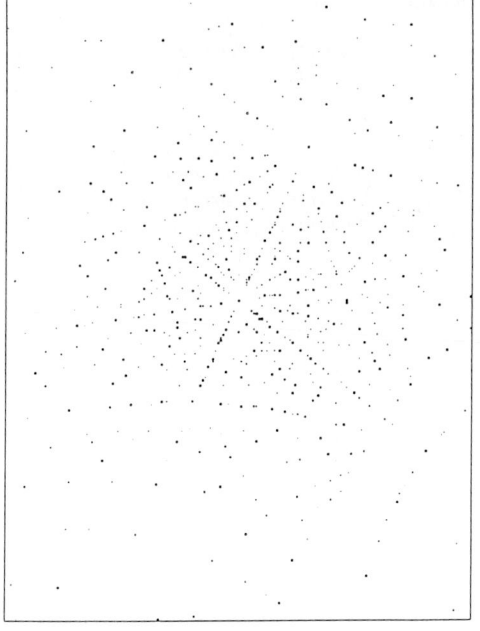

**Figure 5.12b:** *Simulated back-reflection Laue-gram for KTiOPO$_4$, plane $(\bar{1}\,1\,1)$, maximum Miller index 12, detection level 1%, saturation level 15%, sample-to-detector distance 3.1 cm.*

## 5.4 Monoclinic: $(CH_2NH_2COOH)_3 H_2SO_4$

Figure 5.13a shows an experimental Laue-gram of an unknown orientation of a $(CH_2NH_2COOH)_3 H_2SO_4$ (TGS) single crystal (see appendix A2.5). To proceed with the indexing of the Laue-gram, points 1 to 6 are selected, considering spots 1 and 2 as fundamental, according with the rules discussed in section 3.3.1. Then, by introducing into the algorithm the indexing data shown in table 5.10, the solution shown in table 5.11 is suggested.

The simulated Laue-gram shown in figure 5.13b corresponds to the solution presented in table 5.11.

**Table 5.10:** *Inputs for the indexing of the Laue-gram in figure 5.13a.*

|  | x | y |
|---|---|---|
| Maximum Miller index | 2 | |
| Sample-to-detector distance (cm) | 3.0 | |
| Maximum angular tolerance (°) | 0.5 | |
| Maximum index tolerance | 0.07 | |
| Coordinates of fundamental spots (cm) | | |
| Spot 1 | 0.050 (25) | -2.600 (25) |
| Spot 2 | -2.500 (25) | -1.100 (25) |
| Coordinates of the other spots (cm) | | |
| Spot 3 | 0.100 (25) | -1.000 (25) |
| Spot 4 | -2.250 (25) | -0.700 (25) |
| Spot 5 | 1.650 (25) | -2.300 (25) |
| Spot 6 | 3.900 (25) | 1.950 (25) |

**Table 5.11:** *Indexing solutions for the Laue-gram in figure 5.13a with the inputs of table 5.10.*

|  | Solution |
|---|---|
| 1 | $2\ 2\ \bar{1}$ |
| 2 | $1\ 2\ \bar{1}$ |
| 3 | $2\ 3\ \bar{1}$ |
| 4 | $4\ 9\ \bar{4}$ |
| 5 | $3\ 3\ \bar{1}$ |
| 6 | $3\ 7\ 0$ |

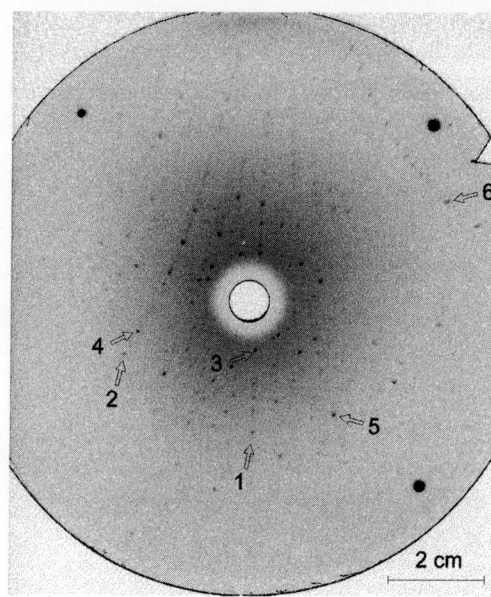

**Figure 5.13a:** *Experimental back-reflection Laue-gram for TGS: Mo tube V=20kV, I=40mA, exposure time 240 min., Kodak AX film, sample-to-detector distance 3.0 cm.*

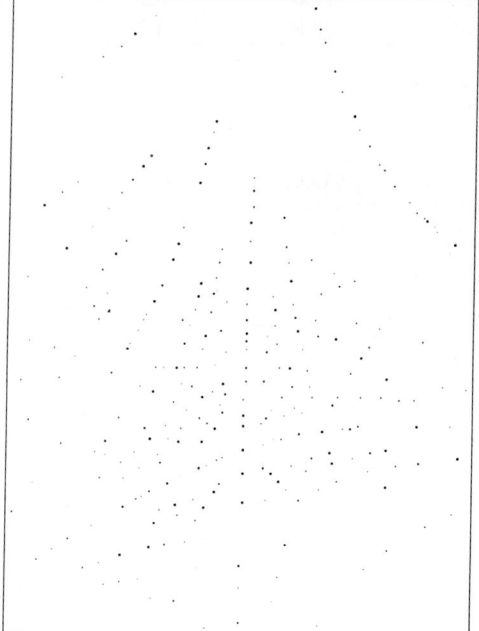

**Figure 5.13b:** *Simulated back-reflection Laue-gram for TGS. Solution in table 5.11; maximum Miller index 12, detection level 0.1%, saturation level 40%, sample-to-detector distance 3.0 cm.*

Four combinations of angles to get the orientations (0 1 0), (1 1 0) and (2 1 $\bar{1}$) are calculated from the solution shown in table 5.11 (table 5.12). The experimental and simulated Laue-grams shown in figures 5.14, 5.15 and 5.16 are obtained by applying the rotation combination α and β. As expected, they correspond to the orientations (0 1 0), (1 1 0) and (2 1 $\bar{1}$), respectively.

**Table 5.12:** *Rotation angles (four combinations) calculated to obtain the orientations (0 1 0), (1 1 0) and (2 1 $\bar{1}$) from the solution in table 5.11.*

|   | $(0\ 1\ 0)$ | $(1\ 1\ 0)$ | $(2\ 1\ \bar{1})$ |
|---|---|---|---|
| α | 2.3° | 38.1° | -0.2° |
| β | 35.9° | -7.7° | -35.7° |
| β | 35.9° | -9.7° | -35.7° |
| α | 1.9° | 37.7° | -0.1° |
| γ | 86.8° | -12.3° | 89.7° |
| α | 36.0° | 38.8° | -35.7° |
| γ | -3.2° | 77.6° | -0.2° |
| β | 36.0° | -38.8° | -35.7° |

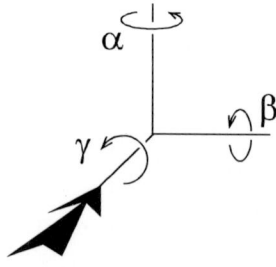

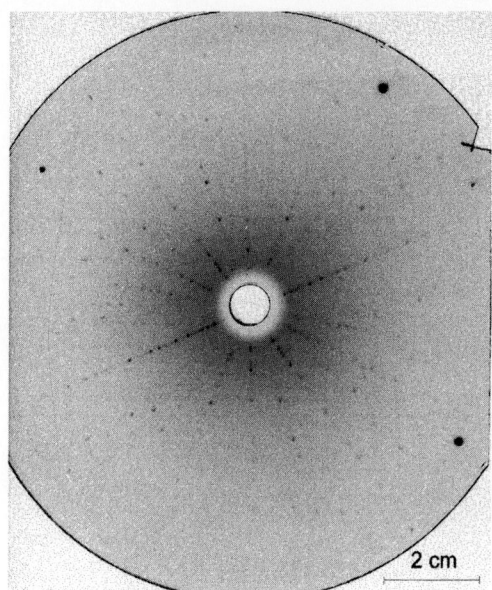

**Figure 5.14a:** *Experimental back-reflection Laue-gram for TGS: plane (0 1 0), Mo tube V=20kV, I=40mA, exposure time 240 min., Kodak AX film, sample-to-detector distance 3.1 cm.*

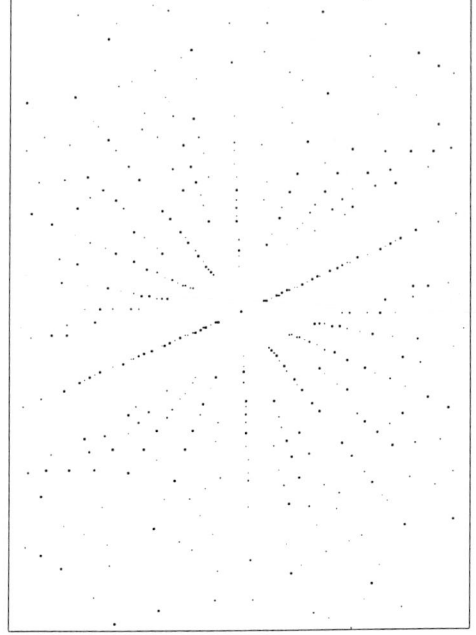

**Figure 5.14b:** *Simulated back-reflection Laue-gram for TGS, plane (0 1 0), maximum Miller index 18, detection level 1%, saturation level 20%, sample-to-detector distance 3.1 cm.*

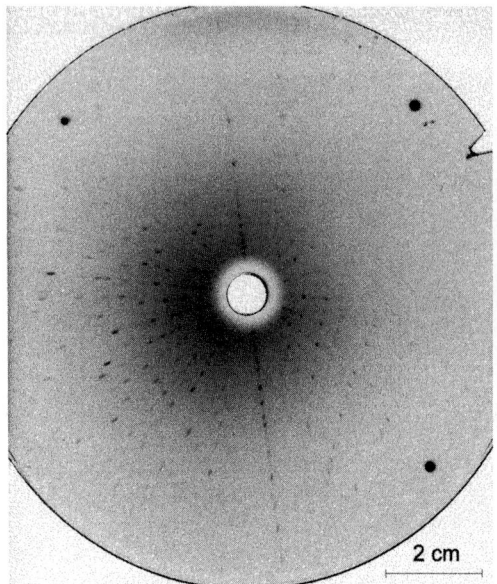

2 cm

**Figure 5.15a:** *Experimental back-reflection Laue-gram for TGS: plane (1 1 0), Mo tube V=20kV, I=40mA, exposure time 240 min., Kodak AX film, sample-to-detector distance 3.1 cm.*

**Figure 5.15b:** *Simulated back-reflection Laue-gram for TGS, plane (1 1 0), maximum Miller index 18, detection level 1%, saturation level 25%, sample-to-detector distance 3.1 cm.*

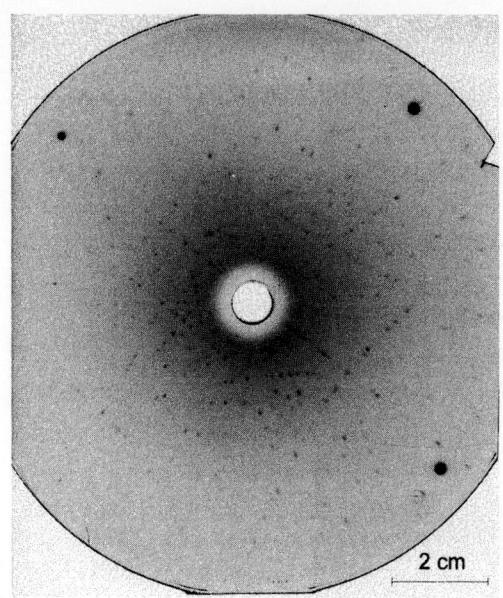

**Figure 5.16a:** *Experimental back-reflection Laue-gram for TGS: plane (2 1 1̄), Mo tube V=20kV, I=40mA, exposure time 240 min., Kodak AX film, sample-to-detector distance 3.1 cm.*

2 cm

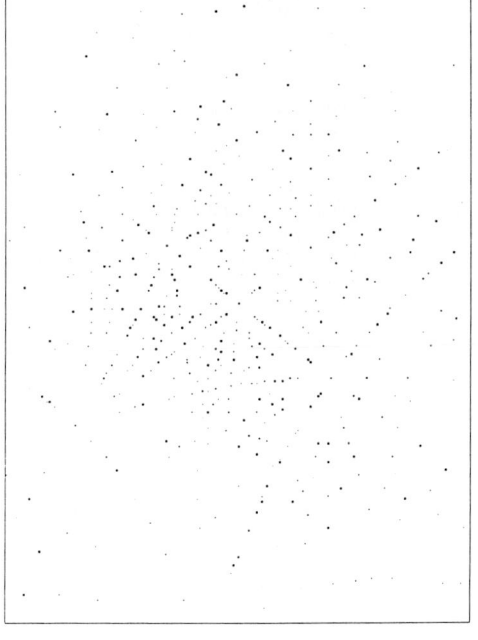

**Figure 5.16b:** *Simulated back-reflection Laue-gram for TGS, plane (2 1 1̄), maximum Miller index 14, detection level 1%, saturation level 20%, sample-to-detector distance 3.1 cm.*

## 5.5 Hexagonal: ZnO

Figure 5.17a shows an experimental Laue-gram of an unknown orientation of a ZnO single crystal (see appendix A2.6). To proceed with the indexing of the Laue-gram, points 1 to 8 are selected, considering spots 1 and 2 as fundamental, according with the rules discussed in section 3.3.1. Then, by introducing into the algorithm the indexing data shown in table 5.13, the 12 solutions shown in table 5.14 are suggested.

The simulated Laue-gram shown in figure 5.17b corresponds to all indexing solutions presented in table 5.14 for the symmetry involved.

It is worth mentioning the shape of the spots shown in the experimental patterns of this material. The enlargement obeys the laminate shape of the sample employed, which determines the geometry of the intersection and, therefore, the aspect of the spot.

**Table 5.13:** *Inputs for the indexing of the Laue-gram in figure 5.17a.*

| | x | y |
|---|---|---|
| Maximum Miller index | 4 | |
| Sample-to-detector distance (cm) | 3.0 | |
| Maximum angular tolerance (°) | 0.3 | |
| Maximum index tolerance | 0.12 | |
| Coordinates of fundamental spots (cm) | | |
| Spot 1 | 2.900 (25) | 1.550 (25) |
| Spot 2 | -0.200 (25) | 2.700 (25) |
| Coordinates of the other spots (cm) | | |
| Spot 3 | -2.900 (25) | -2.750 (25) |
| Spot 4 | 2.700 (25) | -1.550 (25) |
| Spot 5 | 2.200 (25) | 1.000 (25) |
| Spot 6 | -0.150 (25) | 1.950 (25) |
| Spot 7 | 1.750 (25) | 3.150 (25) |
| Spot 8 | -2.050 (25) | -2.000 (25) |

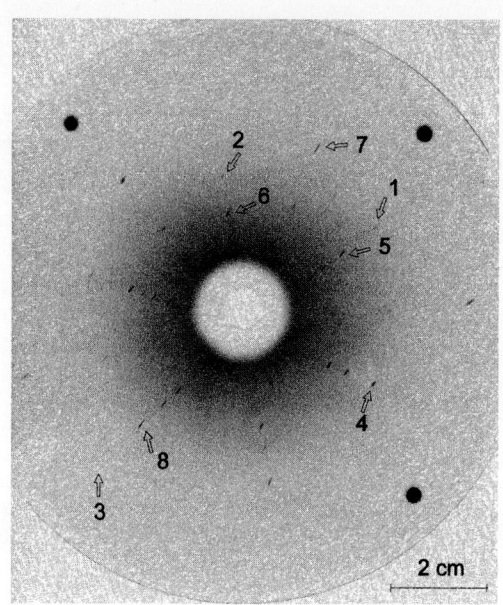

**Figure 5.17a:** *Experimental back-reflection Laue-gram for ZnO: Mo tube V=20kV, I=40mA, exposure time 240 min., Kodak AX film, sample-to-detector distance 3 cm.*

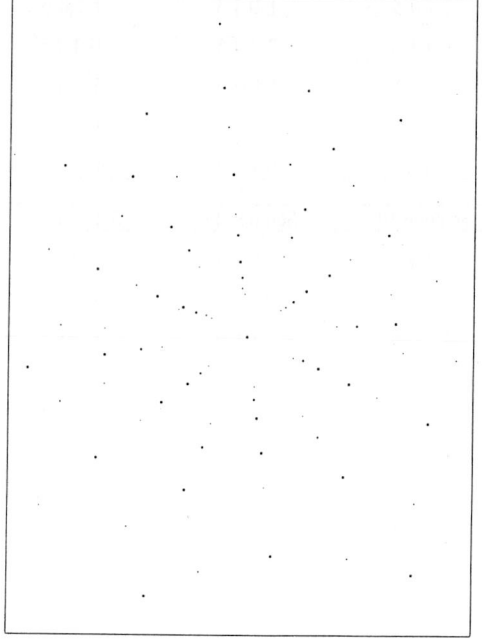

**Figure 5.17b:** *Simulated back-reflection Laue-gram for ZnO. Solution 12 in table 5.14; maximum Miller index 15, detection level 0.01%, saturation level 1.2%, sample-to-detector distance 3 cm.*

**Table 5.14:** *Indexing solutions for the Laue-gram in figure 5.17a with the inputs of table 5.13.*

|  | Solution 1 | Solution 2 | Solution 3 | Solution 4 |
|---|---|---|---|---|
| 1 | $\bar{1}01\bar{4}$ | $\bar{1}014$ | $\bar{1}10\bar{4}$ | $\bar{1}104$ |
| 2 | $0\bar{1}1\bar{4}$ | $\bar{1}10\bar{4}$ | $\bar{1}01\bar{4}$ | $0\bar{1}14$ |
| 3 | $10\bar{1}\bar{4}$ | $10\bar{1}4$ | $\bar{1}\bar{1}0\bar{4}$ | $1\bar{1}04$ |
| 4 | $\bar{1}10\bar{5}$ | $0\bar{1}15$ | $0\bar{1}1\bar{5}$ | $\bar{1}015$ |
| 5 | $\bar{1}01\bar{5}$ | $\bar{1}015$ | $\bar{1}10\bar{5}$ | $\bar{1}105$ |
| 6 | $0\bar{1}1\bar{5}$ | $\bar{1}105$ | $\bar{1}01\bar{5}$ | $0\bar{1}15$ |
| 7 | $\bar{1}\bar{1}2\bar{6}$ | $\bar{2}116$ | $\bar{2}11\bar{6}$ | $\bar{1}2\bar{1}6$ |
| 8 | $10\bar{1}\bar{5}$ | $10\bar{1}5$ | $\bar{1}10\bar{5}$ | $1\bar{1}05$ |

|  | Solution 5 | Solution 6 | Solution 7 | Solution 8 |
|---|---|---|---|---|
| 1 | $0\bar{1}1\bar{4}$ | $0\bar{1}14$ | $0\bar{1}\bar{1}4$ | $0\bar{1}14$ |
| 2 | $\bar{1}10\bar{4}$ | $\bar{1}014$ | $\bar{1}10\bar{4}$ | $10\bar{1}4$ |
| 3 | $01\bar{1}\bar{4}$ | $01\bar{1}4$ | $0\bar{1}1\bar{4}$ | $0\bar{1}14$ |
| 4 | $\bar{1}01\bar{5}$ | $\bar{1}105$ | $10\bar{1}\bar{5}$ | $\bar{1}105$ |
| 5 | $0\bar{1}1\bar{5}$ | $0\bar{1}15$ | $01\bar{1}\bar{5}$ | $0\bar{1}15$ |
| 6 | $\bar{1}10\bar{5}$ | $\bar{1}015$ | $\bar{1}10\bar{5}$ | $10\bar{1}5$ |
| 7 | $1\bar{2}1\bar{6}$ | $\bar{1}\bar{1}26$ | $\bar{1}2\bar{1}6$ | $11\bar{2}6$ |
| 8 | $0\bar{1}1\bar{5}$ | $0\bar{1}15$ | $0\bar{1}1\bar{5}$ | $0\bar{1}15$ |

|  | Solution 9 | Solution 10 | Solution 11 | Solution 12 |
|---|---|---|---|---|
| 1 | $1\bar{1}0\bar{4}$ | $1\bar{1}04$ | $10\bar{1}4$ | $10\bar{1}4$ |
| 2 | $10\bar{1}\bar{4}$ | $0\bar{1}14$ | $0\bar{1}\bar{1}4$ | $\bar{1}104$ |
| 3 | $\bar{1}10\bar{4}$ | $\bar{1}104$ | $\bar{1}01\bar{4}$ | $\bar{1}014$ |
| 4 | $0\bar{1}1\bar{5}$ | $10\bar{1}5$ | $\bar{1}10\bar{5}$ | $0\bar{1}15$ |
| 5 | $1\bar{1}0\bar{5}$ | $1\bar{1}05$ | $10\bar{1}\bar{5}$ | $10\bar{1}5$ |
| 6 | $10\bar{1}\bar{5}$ | $0\bar{1}15$ | $0\bar{1}\bar{1}\bar{5}$ | $\bar{1}105$ |
| 7 | $2\bar{1}\bar{1}\bar{6}$ | $1\bar{2}16$ | $11\bar{2}\bar{6}$ | $2\bar{1}\bar{1}6$ |
| 8 | $\bar{1}10\bar{5}$ | $\bar{1}105$ | $\bar{1}01\bar{5}$ | $\bar{1}015$ |

Four combinations of angles to get the orientations (0 0 0 1), (0 $\bar{1}$ 1 2) and ($\bar{1}$ 1 0 3) are calculated from the 12 solution shown in table 5.14 (table 5.15). The experimental and simulated Laue-grams shown in figures 5.18, 5.19 and 5.20 are obtained by applying the rotation combination α and β. As expected, they correspond to the orientations (0 0 0 1), (0 $\bar{1}$ 1 2) and ($\bar{1}$ 1 0 3), respectively.

**Table 5.15:** *Rotation angles (four combinations) calculated to obtain the orientations (0 0 0 1), (0 $\bar{1}$ 1 2) and ($\bar{1}$ 1 0 3) from the 12 solution in table 5.14.*

|     | (0 0 0 1) | (0 $\bar{1}$ 1 2) | ($\bar{1}$ 1 0 3) |
|-----|-----------|-------------------|-------------------|
| α   | 1.0°      | -38.6°            | 4.8°              |
| β   | -3.5°     | 13.4°             | -35.0°            |
| β   | -3.5°     | 16.9°             | -35.1°            |
| α   | 1.0°      | -37.4°            | -3.9°             |
| γ   | -74.0°    | -20.9°            | -83.2°            |
| α   | 3.6°      | -40.5°            | 35.3°             |
| γ   | 16.0°     | 69.1°             | 6.8°              |
| β   | -3.6°     | 40.5°             | -35.3°            |

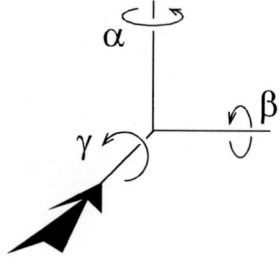

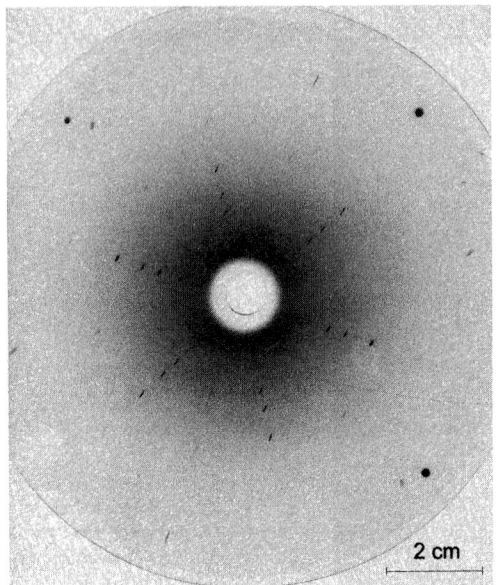

**Figure 5.18a:** *Experimental back-reflection Laue-gram for ZnO: plane (0 0 0 1), Mo tube V=20kV, I=40mA, exposure time 240 min., Kodak AX film, sample-to-detector distance 3 cm.*

**Figure 5.18b:** *Simulated back-reflection Laue-gram for ZnO, plane (0 0 0 1), maximum Miller index 25, detection level 0.01%, saturation level 1.2%, sample-to-detector distance 3 cm.*

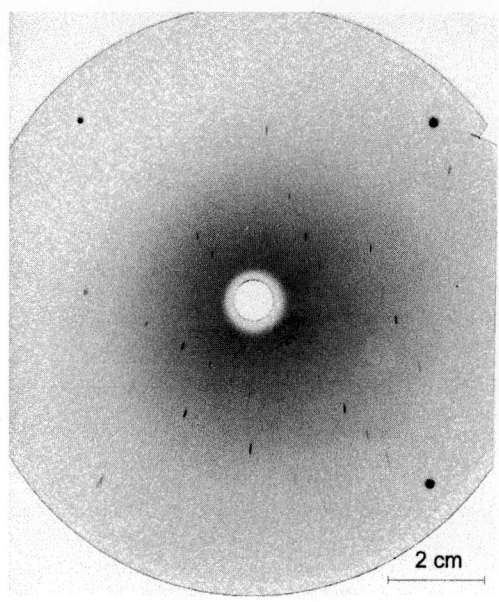

**Figure 5.19a:** *Experimental back-reflection Laue-gram for ZnO: plane (0 1̄ 1 2 ), Mo tube V=20kV, I=40mA, exposure time 240 min., Kodak AX film, sample-to-detector distance 3 cm.*

2 cm

**Figure 5.19b:** *Simulated back-reflection Laue-gram for ZnO, plane (0 1̄ 1 2 ), maximum Miller index 24, detection level 0.01%, saturation level 1.5%, sample-to-detector distance 3 cm.*

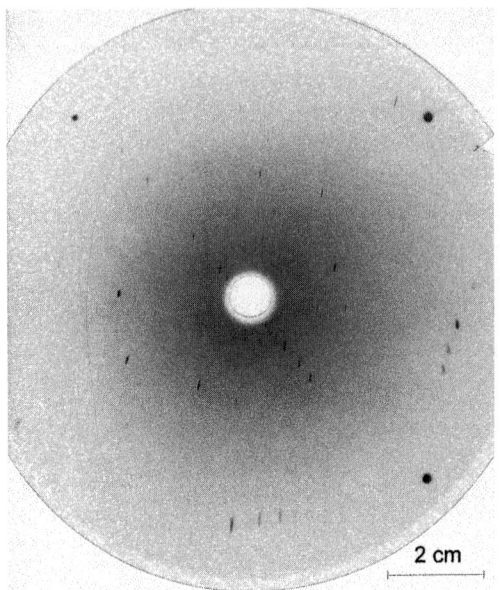

2 cm

**Figure 5.20a:** *Experimental back-reflection Laue-gram for ZnO: plane (1̄ 1 0 3), Mo tube V=20kV, I=40mA, exposure time 240 min., Kodak AX film, sample-to-detector distance 3 cm.*

**Figure 5.20b:** *Simulated back-reflection Laue-gram for ZnO, plane (1̄ 1 0 3), maximum Miller index 20, detection level 0.01%, saturation level 1.5%, sample-to-detector distance 3 cm.*

## 5.6 Rhombohedral: LiNbO₃

Figure 5.21a shows an experimental Laue-gram of an unknown orientation of a LiNbO$_3$ single crystal (see appendix A2.7). The two notations (23.b) applicable to the rhombohedral system are going to be used to proceed with the indexing of the Laue-gram. Points 1r to 5r are selected to be used with the rhombohedral notation, while spots 1r to 3r are selected for the hexagonal notation. Spots 1 and 2 are chosen as fundamental, according with the rules discussed in section 3.3.1. Then, by introducing into the algorithm the indexing data shown in table 5.16 (rhombohedral) or table 5.17 (hexagonal), the 6 solutions shown in table 5.18 or in table 5.19 are suggested for the rhombohedral or the hexagonal notation respectively.

The simulated Laue-gram shown in figure 5.21b corresponds to all indexing solutions presented in tables 5.18 and 5.19 for the symmetry involved.

| | x | y |
|---|---|---|
| Maximum Miller index | 1 | |
| Sample-to-detector distance (cm) | 3.1 | |
| Maximum angular tolerance (°) | 1 | |
| Maximum index tolerance | 0.1 | |
| Coordinates of fundamental spots (cm) | | |
| Spot 1 | -2.800 (25) | -1.050 (25) |
| Spot 2 | 3.550 (25) | -2.350 (25) |
| Coordinates of the other spots (cm) | | |
| Spot 3 | 1.200 (25) | 1.350 (25) |
| Spot 4 | 0.550 (25) | -1.450 (25) |
| Spot 5 | -1.200 (25) | -4.700 (25) |

**Table 5.16:** *Inputs for the indexing of the Laue-gram in figure 5.21a (rhombohedral notation).*

| | x | y |
|---|---|---|
| Maximum Miller index | 1 | |
| Sample-to-detector distance (cm) | 3.1 | |
| Maximum angular tolerance (°) | 1 | |
| Maximum index tolerance | 0.1 | |
| Coordinates of fundamental spots (cm) | | |
| Spot 1 | -2.800 (25) | -1.050 (25) |
| Spot 2 | 3.550 (25) | -2.350 (25) |
| Coordinates of the other spots (cm) | | |
| Spot 3 | -1.200 (25) | -4.700 (25) |

**Table 5.17:** *Inputs for the indexing of the Laue-gram in figure 5.21a (hexagonal notation).*

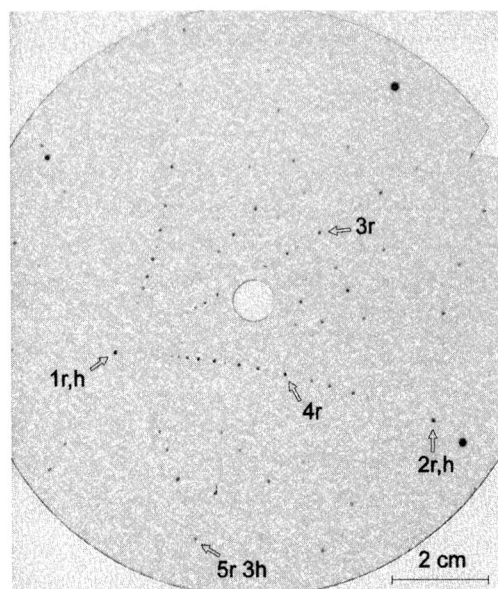

**Figure 5.21a:** *Experimental back-reflection Laue-gram for LiNbO₃: Mo tube V=20kV, I=40mA, exposure time 150 min., Kodak AX film, sample-to-detector distance 3.1 cm.*

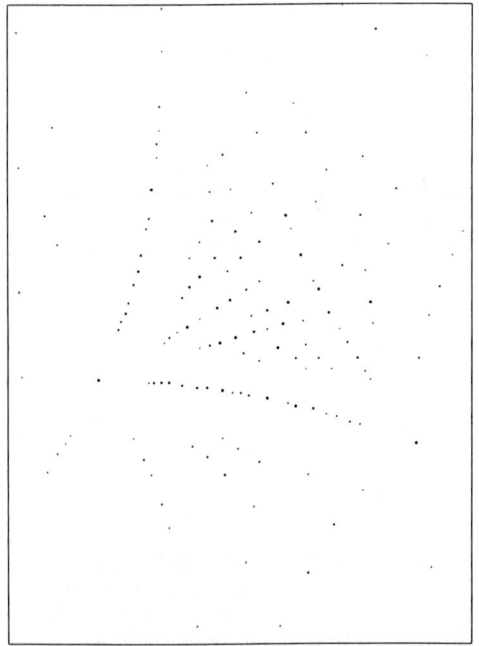

**Figure 5.21b:** *Simulated back-reflection Laue-gram for LiNbO₃. Solution 4 in table 5.19; maximum Miller index 30, detection level 20%, saturation level 60%, sample-to-detector distance 3.1 cm.*

**Table 5.18:** *Indexing solutions for the Laue-gram in figure 5.21a with the inputs of table 5.16.*

|   | Solution 1 | Solution 2 | Solution 3 | Solution 4 |
|---|---|---|---|---|
| 1 | $\bar{1}\bar{1}0$ | $\bar{1}0\bar{1}$ | $0\bar{1}\bar{1}$ | $011$ |
| 2 | $0\bar{1}1$ | $\bar{1}10$ | $10\bar{1}$ | $\bar{1}10$ |
| 3 | $\bar{1}30$ | $\bar{3}0\bar{1}$ | $0\bar{1}3$ | $031$ |
| 4 | $\bar{1}21$ | $\bar{2}1\bar{1}$ | $1\bar{1}2$ | $\bar{1}21$ |
| 5 | $\bar{1}\bar{1}1$ | $\bar{1}\bar{1}\bar{1}$ | $\bar{1}\bar{1}\bar{1}$ | $\bar{1}11$ |

|   | Solution 5 | Solution 6 |
|---|---|---|
| 1 | $101$ | $110$ |
| 2 | $0\bar{1}1$ | $10\bar{1}$ |
| 3 | $103$ | $310$ |
| 4 | $1\bar{1}2$ | $21\bar{1}$ |
| 5 | $1\bar{1}\bar{1}$ | $1\bar{1}\bar{1}$ |

**Table 5.19:** *Indexing solutions for the Laue-gram in figure 5.21a with the inputs of table 5.17.*

|   | Solution 1 | Solution 2 | Solution 3 | Solution 4 |
|---|---|---|---|---|
| 1 | $\bar{1}012$ | $\bar{1}10\bar{2}$ | $0\bar{1}1\bar{2}$ | $01\bar{1}2$ |
| 2 | $\bar{2}110$ | $\bar{2}110$ | $1\bar{2}10$ | $1\bar{1}20$ |
| 3 | $\bar{2}021$ | $\bar{2}20\bar{1}$ | $0\bar{2}2\bar{1}$ | $02\bar{2}1$ |

|   | Solution 5 | Solution 6 |
|---|---|---|
| 1 | $1\bar{1}02$ | $10\bar{1}\bar{2}$ |
| 2 | $1\bar{2}10$ | $11\bar{2}0$ |
| 3 | $2\bar{2}01$ | $20\bar{2}\bar{1}$ |

Four combinations of angles to get the orientations $(2\,\bar{1}\,\bar{1})$, $(1\,1\,1)$ and $(1\,\bar{1}\,0)$ are calculated from the solution shown in table 5.18 (table 5.20), which are equivalent respectively to the orientations $(3\,0\,\bar{3}\,0)$, $(0\,0\,0\,3)$ and $(2\,\bar{1}\,\bar{1}\,0)$ shown in table 5.19 for hexagonal. The experimental and simulated Laue-grams shown in figures 5.22, 5.23 and 5.24 are obtained by applying the rotation combination $\alpha$ and $\beta$. As expected, they correspond to the orientations $(3\,0\,\bar{3}\,0)$, $(0\,0\,0\,3)$ and $(2\,\bar{1}\,\bar{1}\,0)$, respectively. Or to their equivalent orientations for the rhombohedral notation.

**Table 5.20:** *Rotation angles (four combinations) calculated to obtain the orientations $(3\,0\,\bar{3}\,0)$, $(0\,0\,0\,3)$ and $(2\,\bar{1}\,\bar{1}\,0)$ from the solution in table 5.19.*

|          | $(3\,0\,\bar{3}\,0)$ | $(0\,0\,0\,3)$ | $(2\,\bar{1}\,\bar{1}\,0)$ |
|----------|-----------|----------|----------|
| $\alpha$ | 42.7°     | -54.3°   | 65.0°    |
| $\beta$  | 8.1°      | 40.6°    | 29.7°    |
| $\beta$  | 11.0°     | 55.8°    | 53.4°    |
| $\alpha$ | 42.2°     | -38.0°   | 51.9°    |
| $\gamma$ | 11.9°     | -46.6°   | 32.2°    |
| $\alpha$ | 43.3°     | -63.7°   | 68.4°    |
| $\gamma$ | -78.1°    | 43.4°    | -57.8°   |
| $\beta$  | 43.3°     | 63.7°    | 68.4°    |

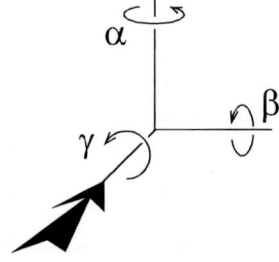

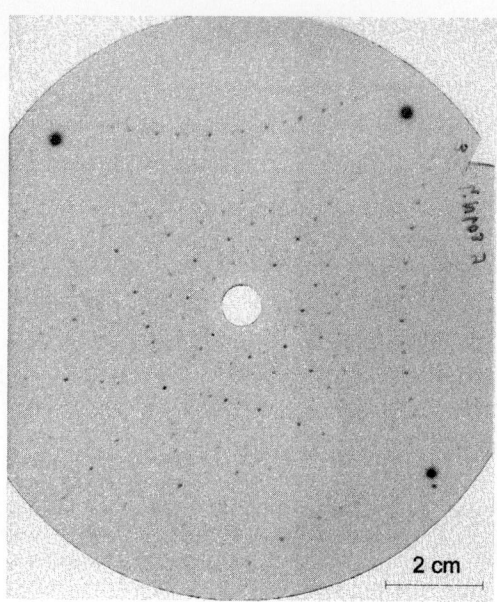

2 cm

**Figure 5.22a:** *Experimental back-reflection Laue-gram for LiNbO₃: hexagonal notation plane (3 0 3̄ 0), Mo tube V=20kV, I=40mA, exposure time 150 min., Kodak AX film, sample-to-detector distance 3.15 cm.*

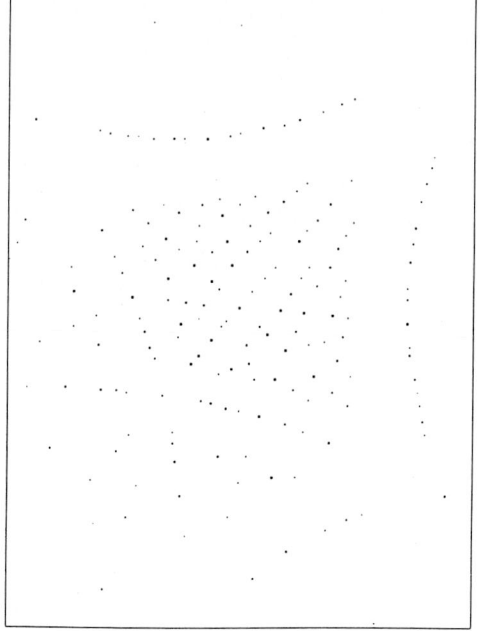

....

**Figure 5.22b:** *Simulated back-reflection Laue-gram for LiNbO₃, hexagonal notation plane (3 0 3̄ 0), maximum Miller index 25, detection level 1%, saturation level 60%, sample-to-detector distance 3.15 cm.*

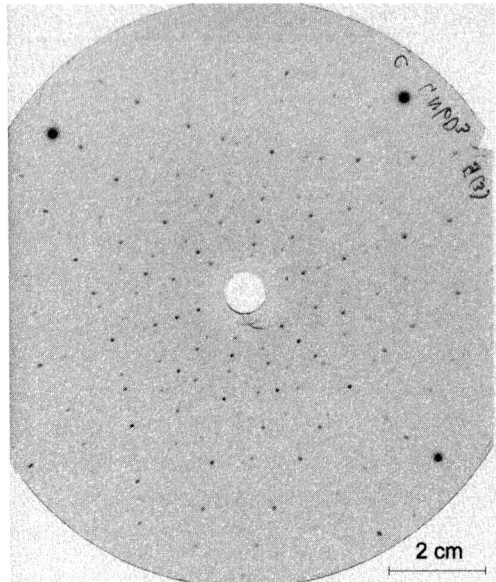

2 cm

**Figure 5.23a:** *Experimental back-reflection Laue-gram for LiNbO₃: hexagonal notation plane (0 0 0 3), Mo tube V=20kV, I=40mA, exposure time 150 min., Kodak AX film, sample-to-detector distance 3.05 cm.*

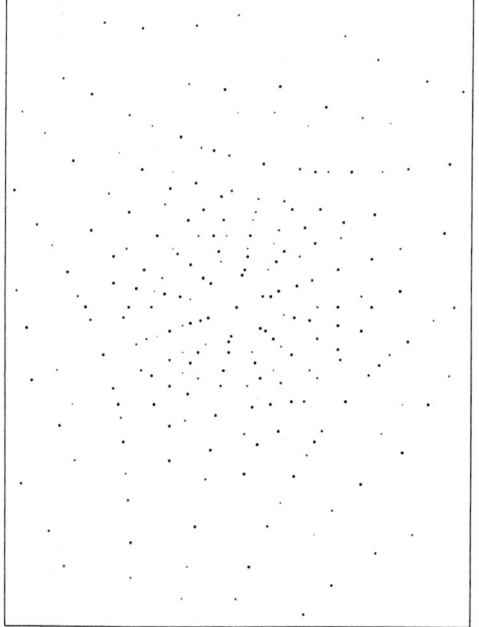

**Figure 5.23b:** *Simulated back-reflection Laue-gram for LiNbO₃, hexagonal notation plane (0 0 0 3), maximum Miller index 40, detection level 2%, saturation level 45%, sample-to-detector distance 3.05 cm.*

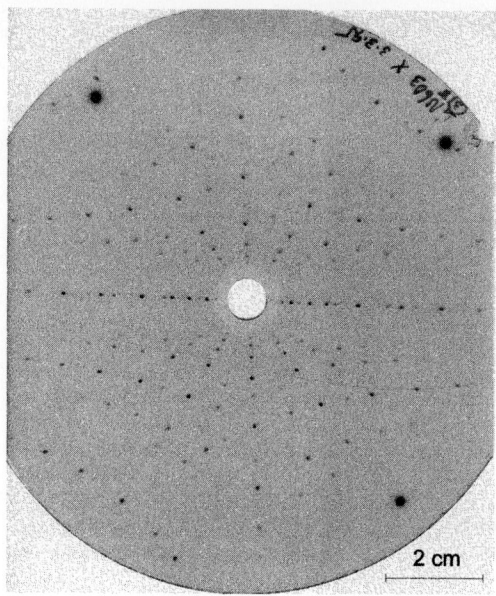

**Figure 5.24a:** *Experimental back-reflection Laue-gram for LiNbO₃: hexagonal notation plane (2 $\bar{1}$ $\bar{1}$ 0), Mo tube V=20kV, I=40mA, exposure time 150 min., Kodak AX film, sample-to-detector distance 3.15 cm.*

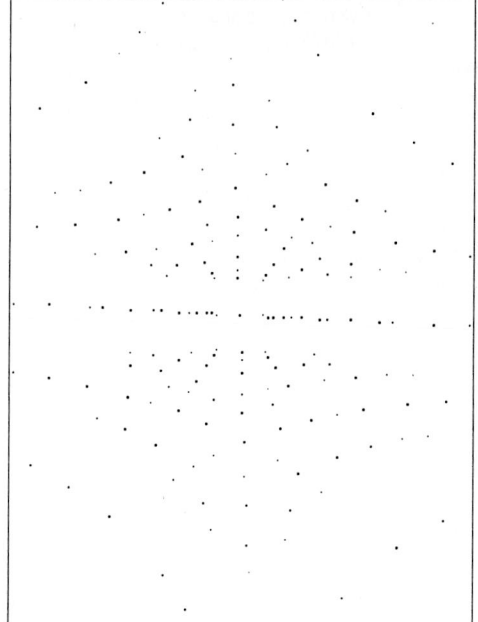

**Figure 5.24b:** *Simulated back-reflection Laue-gram for LiNbO₃, hexagonal notation plane (2 $\bar{1}$ $\bar{1}$ 0), maximum Miller index 25, detection level 8.7%, saturation level 40%, sample-to-detector distance 3.15 cm.*

## 5.7 Triclinic: $Na_2W_4O_{13}$

Figure 5.25a shows an experimental Laue-gram of an unknown orientation of a $Na_2W_4O_{13}$ single crystal (see appendix A2.8). To proceed with the indexing of the Laue-gram, points 1 to 4 are selected, considering spots 1 and 2 as fundamental, according with the rules discussed in section 3.3.1. Then, by introducing into the algorithm the indexing data shown in table 5.21, the solution shown in table 5.22 is suggested.

The simulated Laue-gram shown in figure 5.25b corresponds to the solution presented in table 5.22.

**Table 5.21:** *Inputs for the indexing of the Laue-gram in figure 5.25a.*

| | x | y |
|---|---|---|
| Maximum Miller index | 2 | |
| Sample-to-detector distance (cm) | 3.5 | |
| Maximum angular tolerance (°) | 1 | |
| Maximum index tolerance | 0.06 | |
| Coordinates of fundamental spots (cm) | | |
| Spot 1 | -0.600 (25) | 2.500 (25) |
| Spot 2 | 4.650 (25) | -2.150 (25) |
| Coordinates of the other spots (cm) | | |
| Spot 3 | 1.900 (25) | 0.050 (25) |
| Spot 4 | -2.600 (25) | 0.250 (25) |

**Table 5.22:** *Indexing solutions for the Laue-gram in figure 5.25a with the inputs of table 5.21.*

| | Solution |
|---|---|
| 1 | 1 0 0 |
| 2 | 2 0 $\bar{1}$ |
| 3 | 3 0 $\bar{1}$ |
| 4 | 6 1 0 |

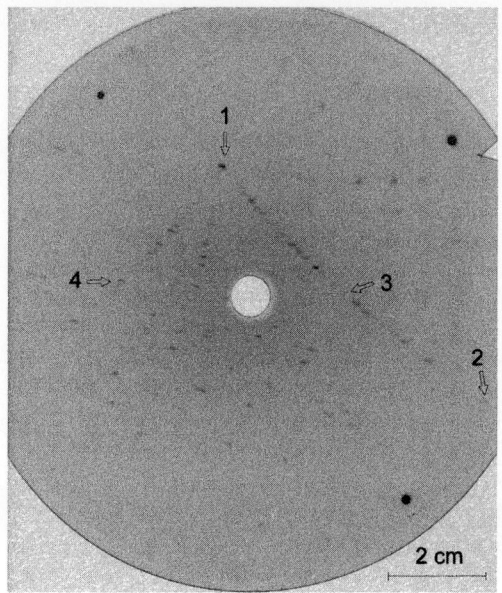

**Figure 5.25a:** *Experimental back-reflection Laue-gram for $Na_2W_4O_{13}$: Mo tube V=20kV, I=40mA, exposure time 180 min., Kodak AX film, sample-to-detector distance 3.5 cm.*

**Figure 5.25b:** *Simulated back-reflection Laue-gram for $Na_2W_4O_{13}$. Solution in table 5.22; maximum Miller index 15, detection level 1%, saturation level 10%, sample-to-detector distance 3.5 cm.*

Four combinations of angles to get the orientations $(1\,0\,0)$, $(1\,0\,1)$ and $(2\,0\,\bar{1})$ are calculated from the solution shown in table 5.22 (table 5.23). The experimental and simulated Laue-grams shown in figures 5.26, 5.27 and 5.28 are obtained by applying the rotation combination $\alpha$ and $\beta$. As expected, they correspond to the orientations $(1\,0\,0)$, $(1\,0\,1)$ and $(2\,0\,\bar{1})$, respectively.

**Table 5.23:** *Rotation angles (four combinations) calculated to obtain the orientations $(1\,0\,0)$, $(1\,0\,1)$ and $(2\,0\,\bar{1})$ from the solution in table 5.22.*

|        | $(1\,0\,0)$ | $(1\,0\,1)$ | $(2\,0\,\bar{1})$ |
|--------|-------------|-------------|-------------------|
| $\alpha$ | -4.4°     | -28.9°      | 25.8°             |
| $\beta$  | 17.6°     | 34.4°       | -11.6°            |
| $\beta$  | 17.7°     | 38.1°       | -12.8°            |
| $\alpha$ | -4.2°     | -23.5°      | 25.2°             |
| $\gamma$ | -76.5°    | -54.8°      | -25.2°            |
| $\alpha$ | -18.1°    | -43.8°      | 28.1°             |
| $\gamma$ | 13.5°     | 35.2°       | 64.7°             |
| $\beta$  | 18.1°     | 43.8°       | -28.1°            |

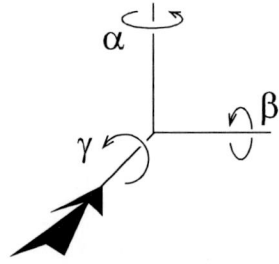

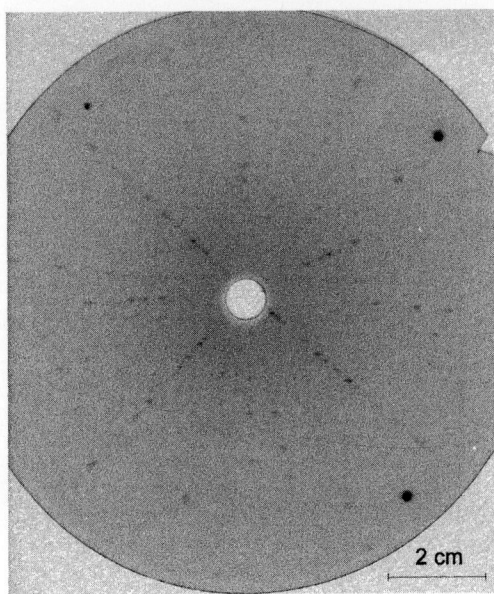

**Figure 5.26a:** *Experimental back-reflection Laue-gram for $Na_2W_4O_{13}$: plane (1 0 0), Mo tube V=20kV, I=40mA, exposure time 180 min., Kodak AX film, sample-to-detector distance 3.5 cm.*

2 cm

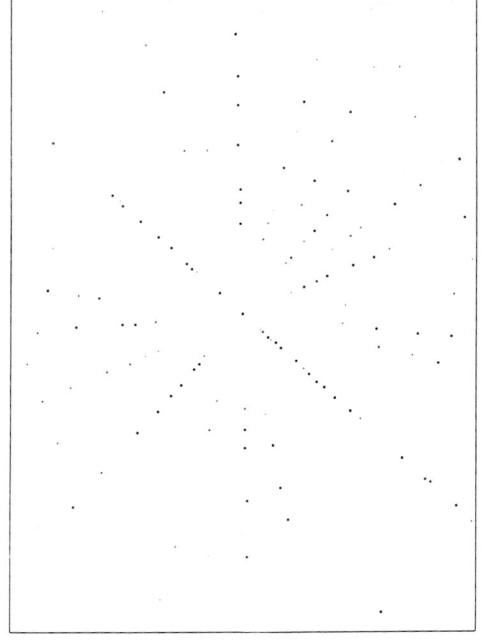

**Figure 5.26b:** *Simulated back-reflection Laue-gram for $Na_2W_4O_{13}$, plane (1 0 0), maximum Miller index 20, detection level 2%, saturation level 10%, sample-to-detector distance 3.5 cm.*

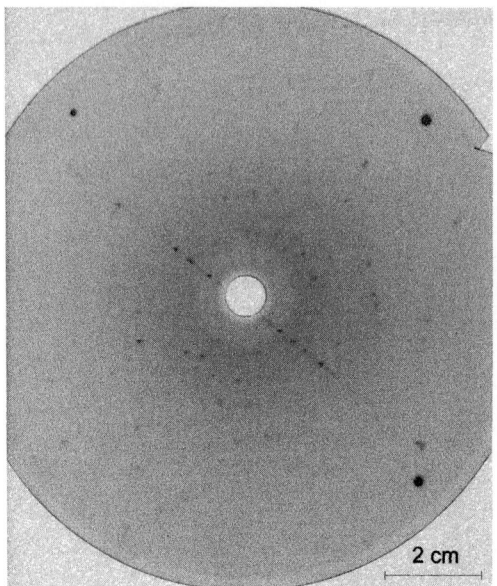

2 cm

**Figure 5.27a:** *Experimental back-reflection Laue-gram for $Na_2W_4O_{13}$: plane (1 0 1), Mo tube V=20kV, I=40mA, exposure time 180 min., Kodak AX film, sample-to-detector distance 3.4 cm.*

**Figure 5.27b:** *Simulated back-reflection Laue-gram for $Na_2W_4O_{13}$, plane (1 0 1), maximum Miller index 18, detection level 1%, saturation level 15%, sample-to-detector distance 3.4 cm.*

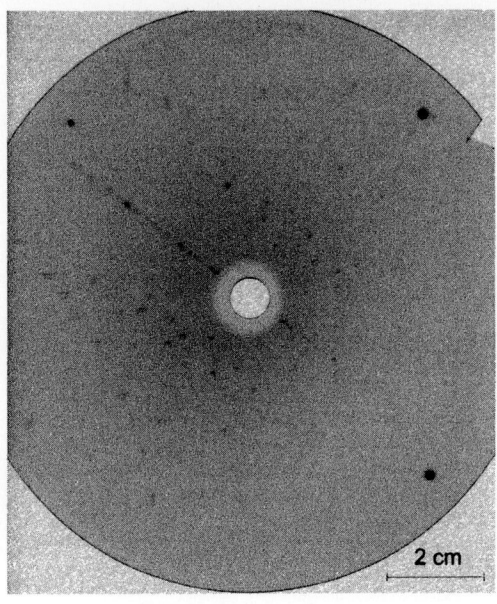

**Figure 5.28a:** *Experimental back-reflection Laue-gram for Na₂W₄O₁₃: plane (2 0 1̄), Mo tube V=20kV, I=40mA, exposure time 180 min., Kodak AX film, sample-to-detector distance 3.25 cm.*

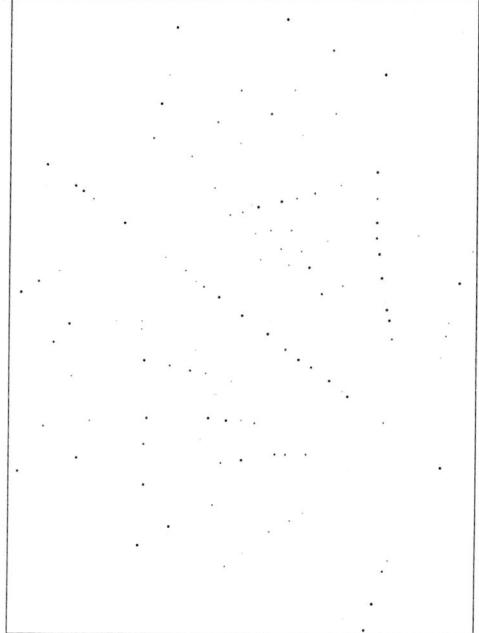

**Figure 5.28b:** *Simulated back-reflection Laue-gram for Na₂W₄O₁₃, plane (2 0 1̄), maximum Miller index 18, detection level 1%, saturation level 12%, sample-to-detector distance 3.25 cm.*

# APPENDICES

## A1 The 230 Space Groups

| Space Group | Schoenflies notation | Mauguin-Hermann notation |
|---|---|---|
| 1 | $C_1^1$ | $P1$ |
| 2 | $C_i^1$ | $P\bar{1}$ |
| 3 | $C_2^1$ | $P2$ |
| 4 | $C_2^2$ | $P2_1$ |
| 5 | $C_2^3$ | $C2$ |
| 6 | $C_s^1$ | $Pm$ |
| 7 | $C_s^2$ | $Pc$ |
| 8 | $C_s^3$ | $Cm$ |
| 9 | $C_s^4$ | $Cc$ |
| 10 | $C_{2h}^1$ | $P2/m$ |
| 11 | $C_{2h}^2$ | $P2_1/m$ |
| 12 | $C_{2h}^3$ | $C2/m$ |
| 13 | $C_{2h}^4$ | $P2/c$ |
| 14 | $C_{2h}^5$ | $P2_1/c$ |
| 15 | $C_{2h}^6$ | $C2/c$ |
| 16 | $D_2^1$ | $P222$ |
| 17 | $D_2^2$ | $P222_1$ |
| 18 | $D_2^3$ | $P2_12_12$ |
| 19 | $D_2^4$ | $P2_12_12_1$ |
| 20 | $D_2^5$ | $C222_1$ |
| 21 | $D_2^6$ | $C222$ |
| 22 | $D_2^7$ | $F222$ |
| 23 | $D_2^8$ | $I222$ |
| 24 | $D_2^9$ | $I2_12_12_1$ |
| 25 | $C_{2v}^1$ | $Pmm2$ |

| Space Group | Schoenflies notation | Mauguin-Hermann notation |
|:---:|:---:|:---:|
| 26 | $C_{2v}^2$ | $Pmc2_1$ |
| 27 | $C_{2v}^3$ | $Pcc2$ |
| 28 | $C_{2v}^4$ | $Pma2$ |
| 29 | $C_{2v}^5$ | $Pca2_1$ |
| 30 | $C_{2v}^6$ | $Pnc2$ |
| 31 | $C_{2v}^7$ | $Pmn2_1$ |
| 32 | $C_{2v}^8$ | $Pba2$ |
| 33 | $C_{2v}^9$ | $Pna2_1$ |
| 34 | $C_{2v}^{10}$ | $Pnn2$ |
| 35 | $C_{2v}^{11}$ | $Cmm2$ |
| 36 | $C_{2v}^{12}$ | $Cmc2_1$ |
| 37 | $C_{2v}^{13}$ | $Ccc2$ |
| 38 | $C_{2v}^{14}$ | $Amm2$ |
| 39 | $C_{2v}^{15}$ | $Abm2$ |
| 40 | $C_{2v}^{16}$ | $Ama2$ |
| 41 | $C_{2v}^{17}$ | $Aba2$ |
| 42 | $C_{2v}^{18}$ | $Fmm2$ |
| 43 | $C_{2v}^{19}$ | $Fdd2$ |
| 44 | $C_{2v}^{20}$ | $Imm2$ |
| 45 | $C_{2v}^{21}$ | $Iba2$ |
| 46 | $C_{2v}^{22}$ | $Ima2$ |
| 47 | $D_{2h}^1$ | $Pmmm$ |
| 48 | $D_{2h}^2$ | $Pnnn$ |
| 49 | $D_{2h}^3$ | $Pccm$ |
| 50 | $D_{2h}^4$ | $Pban$ |
| 51 | $D_{2h}^5$ | $Pmma$ |
| 52 | $D_{2h}^6$ | $Pnna$ |
| 53 | $D_{2h}^7$ | $Pmma$ |
| 54 | $D_{2h}^8$ | $Pcca$ |

| Space Group | Schoenflies notation | Mauguin-Hermann notation |
|---|---|---|
| 55 | $D_{2h}^9$ | *Pbam* |
| 56 | $D_{2h}^{10}$ | *Pccn* |
| 57 | $D_{2h}^{11}$ | *Pbcm* |
| 58 | $D_{2h}^{12}$ | *Pnnm* |
| 59 | $D_{2h}^{13}$ | *Pmmn* |
| 60 | $D_{2h}^{14}$ | *Pbcn* |
| 61 | $D_{2h}^{15}$ | *Pbca* |
| 62 | $D_{2h}^{16}$ | *Pnma* |
| 63 | $D_{2h}^{17}$ | *Cmcm* |
| 64 | $D_{2h}^{18}$ | *Cmca* |
| 65 | $D_{2h}^{19}$ | *Cmmm* |
| 66 | $D_{2h}^{20}$ | *Cccm* |
| 67 | $D_{2h}^{21}$ | *Cmma* |
| 68 | $D_{2h}^{22}$ | *Ccca* |
| 69 | $D_{2h}^{23}$ | *Fmmm* |
| 70 | $D_{2h}^{24}$ | *Fddd* |
| 71 | $D_{2h}^{25}$ | *Immm* |
| 72 | $D_{2h}^{26}$ | *Ibam* |
| 73 | $D_{2h}^{27}$ | *Ibca* |
| 74 | $D_{2h}^{28}$ | *Imma* |
| 75 | $C_4^1$ | *P4* |
| 76 | $C_4^2$ | *P4$_1$* |
| 77 | $C_4^3$ | *P4$_2$* |
| 78 | $C_4^4$ | *P4$_3$* |
| 79 | $C_4^5$ | *I4* |
| 80 | $C_4^6$ | *I4$_1$* |
| 81 | $S_4^1$ | *P$\bar{4}$* |
| 82 | $S_4^2$ | *I$\bar{4}$* |
| 83 | $C_{4h}^1$ | *P4 / m* |

| Space Group | Schoenflies notation | Mauguin-Hermann notation |
|---|---|---|
| 84 | $C_{4h}^2$ | $P4_2/m$ |
| 85 | $C_{4h}^3$ | $P4/n$ |
| 86 | $C_{4h}^4$ | $P4_2/n$ |
| 87 | $C_{4h}^5$ | $I4/m$ |
| 88 | $C_{4h}^6$ | $I4_1/a$ |
| 89 | $D_4^1$ | $P422$ |
| 90 | $D_4^2$ | $P42_12$ |
| 91 | $D_4^3$ | $P4_122$ |
| 92 | $D_4^4$ | $P4_12_12$ |
| 93 | $D_4^5$ | $P4_222$ |
| 94 | $D_4^6$ | $P4_22_12$ |
| 95 | $D_4^7$ | $P4_322$ |
| 96 | $D_4^8$ | $P4_32_12$ |
| 97 | $D_4^9$ | $I422$ |
| 98 | $D_4^{10}$ | $I4_122$ |
| 99 | $C_{4v}^1$ | $P4mm$ |
| 100 | $C_{4v}^2$ | $P4bm$ |
| 101 | $C_{4v}^3$ | $P4_2cm$ |
| 102 | $C_{4v}^4$ | $P4_2nm$ |
| 103 | $C_{4v}^5$ | $P4cc$ |
| 104 | $C_{4v}^6$ | $P4nc$ |
| 105 | $C_{4v}^7$ | $P4_2mc$ |
| 106 | $C_{4v}^8$ | $P4_2bc$ |
| 107 | $C_{4v}^9$ | $I4mm$ |
| 108 | $C_{4v}^{10}$ | $I4cm$ |
| 109 | $C_{4v}^{11}$ | $I4_1md$ |
| 110 | $C_{4v}^{12}$ | $I4_1cd$ |
| 111 | $D_{2d}^1$ | $P\bar{4}2m$ |
| 112 | $D_{2d}^2$ | $P\bar{4}2c$ |

| Space Group | Schoenflies notation | Mauguin-Hermann notation |
|:---:|:---:|:---:|
| 113 | $D_{2d}^3$ | $P\bar{4}2_1m$ |
| 114 | $D_{2d}^4$ | $P\bar{4}2_1c$ |
| 115 | $D_{2d}^5$ | $P\bar{4}m2$ |
| 116 | $D_{2d}^6$ | $P\bar{4}c2$ |
| 117 | $D_{2d}^7$ | $P\bar{4}b2$ |
| 118 | $D_{2d}^8$ | $P\bar{4}n2$ |
| 119 | $D_{2d}^9$ | $I\bar{4}m2$ |
| 120 | $D_{2d}^{10}$ | $I\bar{4}c2$ |
| 121 | $D_{2d}^{11}$ | $I\bar{4}2m$ |
| 122 | $D_{2d}^{12}$ | $I\bar{4}2d$ |
| 123 | $D_{4h}^1$ | $P4/mmm$ |
| 124 | $D_{4h}^2$ | $P4/mcc$ |
| 125 | $D_{4h}^3$ | $P4/nbm$ |
| 126 | $D_{4h}^4$ | $P4/nnc$ |
| 127 | $D_{4h}^5$ | $P4/mbm$ |
| 128 | $D_{4h}^6$ | $P4/mnc$ |
| 129 | $D_{4h}^7$ | $P4/nmm$ |
| 130 | $D_{4h}^8$ | $P4/ncc$ |
| 131 | $D_{4h}^9$ | $P4_2/mmc$ |
| 132 | $D_{4h}^{10}$ | $P4_2/mcm$ |
| 133 | $D_{4h}^{11}$ | $P4_2/nbc$ |
| 134 | $D_{4h}^{12}$ | $P4_2/nnm$ |
| 135 | $D_{4h}^{13}$ | $P4_2/mbc$ |
| 136 | $D_{4h}^{14}$ | $P4_2/mnm$ |
| 137 | $D_{4h}^{15}$ | $P4_2/nmc$ |
| 138 | $D_{4h}^{16}$ | $P4_2/ncm$ |
| 139 | $D_{4h}^{17}$ | $I4/mmm$ |
| 140 | $D_{4h}^{18}$ | $I4/mcm$ |
| 141 | $D_{4h}^{19}$ | $I4_1/amd$ |

| Space Group | Schoenflies notation | Mauguin-Hermann notation |
|:---:|:---:|:---:|
| 142 | $D_{4h}^{20}$ | $I4_1/acd$ |
| 143 | $C_3^1$ | $P3$ |
| 144 | $C_3^2$ | $P3_1$ |
| 145 | $C_3^3$ | $P3_2$ |
| 146 | $C_3^4$ | $R3$ |
| 147 | $C_{3i}^1$ | $P\bar{3}$ |
| 148 | $C_{3i}^2$ | $R\bar{3}$ |
| 149 | $D_3^1$ | $P312$ |
| 150 | $D_3^2$ | $P321$ |
| 151 | $D_3^3$ | $P3_112$ |
| 152 | $D_3^4$ | $P3_121$ |
| 153 | $D_3^5$ | $P3_212$ |
| 154 | $D_3^6$ | $P3_221$ |
| 155 | $D_3^7$ | $R32$ |
| 156 | $C_{3v}^1$ | $P3m1$ |
| 157 | $C_{3v}^2$ | $P31m$ |
| 158 | $C_{3v}^3$ | $P3c1$ |
| 159 | $C_{3v}^4$ | $P31c$ |
| 160 | $C_{3v}^5$ | $R3m$ |
| 161 | $C_{3v}^6$ | $R3c$ |
| 162 | $D_{3d}^1$ | $P\bar{3}1m$ |
| 163 | $D_{3d}^2$ | $P\bar{3}1c$ |
| 164 | $D_{3d}^3$ | $P\bar{3}m1$ |
| 165 | $D_{3d}^4$ | $P\bar{3}c1$ |
| 166 | $D_{3d}^5$ | $R\bar{3}m$ |
| 167 | $D_{3d}^6$ | $R\bar{3}c$ |
| 168 | $C_6^1$ | $P6$ |
| 169 | $C_6^2$ | $P6_1$ |
| 170 | $C_6^3$ | $P6_5$ |

| Space Group | Schoenflies notation | Mauguin-Hermann notation |
|:---:|:---:|:---:|
| 171 | $C_6^4$ | $P6_2$ |
| 172 | $C_6^5$ | $P6_4$ |
| 173 | $C_6^6$ | $P6_3$ |
| 174 | $C_{3h}^1$ | $P\bar{6}$ |
| 175 | $C_{6h}^1$ | $P6/m$ |
| 176 | $C_{6h}^2$ | $P6_3/m$ |
| 177 | $D_6^1$ | $P622$ |
| 178 | $D_6^2$ | $P6_122$ |
| 179 | $D_6^3$ | $P6_522$ |
| 180 | $D_6^4$ | $P6_222$ |
| 181 | $D_6^5$ | $P6_422$ |
| 182 | $D_6^6$ | $P6_322$ |
| 183 | $C_{6v}^1$ | $P6mm$ |
| 184 | $C_{6v}^2$ | $P6cc$ |
| 185 | $C_{6v}^3$ | $P6_3cm$ |
| 186 | $C_{6v}^4$ | $P6_3mc$ |
| 187 | $D_{3h}^1$ | $P\bar{6}m2$ |
| 188 | $D_{3h}^2$ | $P\bar{6}c2$ |
| 189 | $D_{3h}^3$ | $P\bar{6}2m$ |
| 190 | $D_{3h}^4$ | $P\bar{6}2c$ |
| 191 | $D_{6h}^1$ | $P6/mmm$ |
| 192 | $D_{6h}^2$ | $P6/mcc$ |
| 193 | $D_{6h}^3$ | $P6_3/mcm$ |
| 194 | $D_{6h}^4$ | $P6_3/mmc$ |
| 195 | $T^1$ | $P23$ |
| 196 | $T^2$ | $F23$ |
| 197 | $T^3$ | $I23$ |
| 198 | $T^4$ | $P2_13$ |
| 199 | $T^5$ | $I2_13$ |
| 200 | $T_h^1$ | $Pm\bar{3}$ |

| Space Group | Schoenflies notation | Mauguin-Hermann notation |
|:---:|:---:|:---:|
| 201 | $T_h^2$ | $Pn\bar{3}$ |
| 202 | $T_h^3$ | $Fm\bar{3}$ |
| 203 | $T_h^4$ | $Fd\bar{3}$ |
| 204 | $T_h^5$ | $Im\bar{3}$ |
| 205 | $T_h^6$ | $Pa\bar{3}$ |
| 206 | $T_h^7$ | $Ia\bar{3}$ |
| 207 | $O^1$ | $P432$ |
| 208 | $O^2$ | $P4_232$ |
| 209 | $O^3$ | $F432$ |
| 210 | $O^4$ | $F4_132$ |
| 211 | $O^5$ | $I432$ |
| 212 | $O^6$ | $P4_332$ |
| 213 | $O^7$ | $P4_132$ |
| 214 | $O^8$ | $I4_132$ |
| 215 | $T_d^1$ | $P\bar{4}3m$ |
| 216 | $T_d^2$ | $F\bar{4}3m$ |
| 217 | $T_d^3$ | $I\bar{4}3m$ |
| 218 | $T_d^4$ | $P\bar{4}3n$ |
| 219 | $T_d^5$ | $F\bar{4}3c$ |
| 220 | $T_d^6$ | $I\bar{4}3d$ |
| 221 | $O_h^1$ | $Pm\bar{3}m$ |
| 222 | $O_h^2$ | $Pn\bar{3}n$ |
| 223 | $O_h^3$ | $Pm\bar{3}n$ |
| 224 | $O_h^4$ | $Pn\bar{3}m$ |
| 225 | $O_h^5$ | $Fm\bar{3}m$ |
| 226 | $O_h^6$ | $Fm\bar{3}c$ |
| 227 | $O_h^7$ | $Fd\bar{3}m$ |
| 228 | $O_h^8$ | $Fd\bar{3}c$ |
| 229 | $O_h^9$ | $Im\bar{3}m$ |
| 230 | $O_h^{10}$ | $Ia\bar{3}d$ |

## A2 Crystal structures data

### A2.1 $Bi_{12}GeO_{20}$

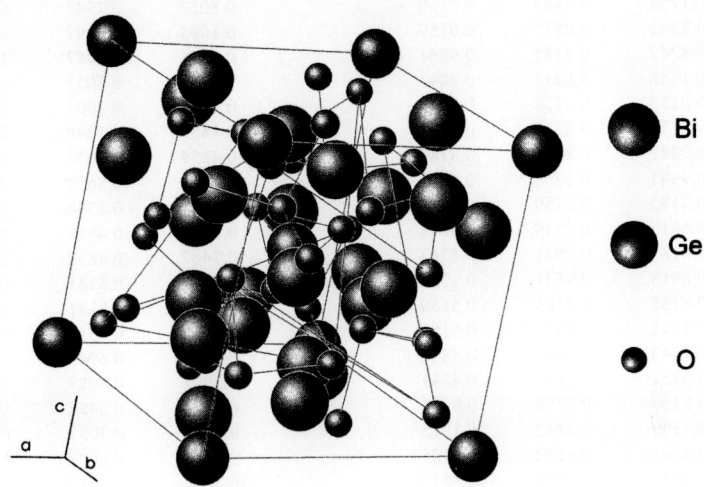

**Figure A2.1** *The $Bi_{12}GeO_{20}$ structure (45).*

| Atomic number / weight of components: | Bi: 83 / 208.98037 |
| | Ge: 32 / 72.61 |
| | O: 8 / 15.994 |

| | |
|---|---|
| Formulas per unit cell: | 2 |
| Crystal system: | Cubic |
| Space group: | *I*23 |
| Unit cell parameter: | a: 10.153Å |
| Atomic relative coordinates (46): | table A2.1 |

**Table A2.1** *Atomic relative coordinates of $Bi_{12}GeO_{20}$.*

| | x | y | z | | x | y | z |
|---|---|---|---|---|---|---|---|
| Bi | 0.1758 | 0.3185 | 0.0159 | | 0.8057 | 0.1943 | 0.8057 |
| | 0.8242 | 0.6815 | 0.0159 | | 0.1003 | 0.8997 | 0.1003 |
| | 0.8242 | 0.3185 | 0.9841 | | 0.1348 | 0.7487 | 0.5141 |
| | 0.1758 | 0.6815 | 0.9841 | | 0.1943 | 0.8057 | 0.8057 |
| | 0.0159 | 0.1758 | 0.3185 | | 0.8997 | 0.1003 | 0.1003 |
| | 0.0159 | 0.8242 | 0.6815 | | 0.4859 | 0.1348 | 0.2513 |
| | 0.9841 | 0.8242 | 0.3185 | | 0.4859 | 0.8652 | 0.7487 |
| | 0.9841 | 0.1758 | 0.6815 | | 0.5141 | 0.8652 | 0.2513 |
| | 0.3185 | 0.0159 | 0.1758 | | 0.5141 | 0.1348 | 0.7487 |
| | 0.6815 | 0.0159 | 0.8242 | | 0.2513 | 0.4859 | 0.1348 |
| | 0.3185 | 0.9841 | 0.8242 | | 0.7487 | 0.4859 | 0.8652 |
| | 0.6815 | 0.9841 | 0.1758 | | 0.2513 | 0.5141 | 0.8652 |
| | 0.6758 | 0.8185 | 0.5159 | | 0.7487 | 0.5141 | 0.1348 |
| | 0.3242 | 0.1815 | 0.5159 | | 0.6348 | 0.7513 | 0.9859 |
| | 0.3242 | 0.8185 | 0.4841 | | 0.6943 | 0.6943 | 0.6943 |
| | 0.6758 | 0.1815 | 0.4841 | | 0.3997 | 0.3997 | 0.3997 |
| | 0.5159 | 0.6758 | 0.8185 | | 0.3652 | 0.2487 | 0.9859 |
| | 0.5159 | 0.3242 | 0.1815 | | 0.3057 | 0.3057 | 0.6943 |
| | 0.4841 | 0.3242 | 0.8185 | | 0.6003 | 0.6003 | 0.3997 |
| | 0.4841 | 0.6758 | 0.1815 | | 0.3652 | 0.7513 | 0.0141 |
| | 0.8185 | 0.5159 | 0.6758 | | 0.3057 | 0.6943 | 0.3057 |
| | 0.1815 | 0.5159 | 0.3242 | | 0.6003 | 0.3997 | 0.6003 |
| | 0.8185 | 0.4841 | 0.3242 | | 0.6348 | 0.2487 | 0.0141 |
| | 0.1815 | 0.4841 | 0.6758 | | 0.6943 | 0.3057 | 0.3057 |
| Ge | 0.0000 | 0.0000 | 0.0000 | | 0.3997 | 0.6003 | 0.6003 |
| | 0.5000 | 0.5000 | 0.5000 | | 0.9859 | 0.6348 | 0.7513 |
| O | 0.1348 | 0.2513 | 0.4859 | | 0.9859 | 0.3652 | 0.2487 |
| | 0.1943 | 0.1943 | 0.1943 | | 0.0141 | 0.3652 | 0.7513 |
| | 0.8997 | 0.8997 | 0.8997 | | 0.0141 | 0.6348 | 0.2487 |
| | 0.8652 | 0.7487 | 0.4859 | | 0.7513 | 0.9859 | 0.6348 |
| | 0.8057 | 0.8057 | 0.1943 | | 0.2487 | 0.9859 | 0.3652 |
| | 0.1003 | 0.1003 | 0.8997 | | 0.7513 | 0.0141 | 0.3652 |
| | 0.8652 | 0.2513 | 0.5141 | | 0.2487 | 0.0141 | 0.6348 |

*A2.2 GaSb*

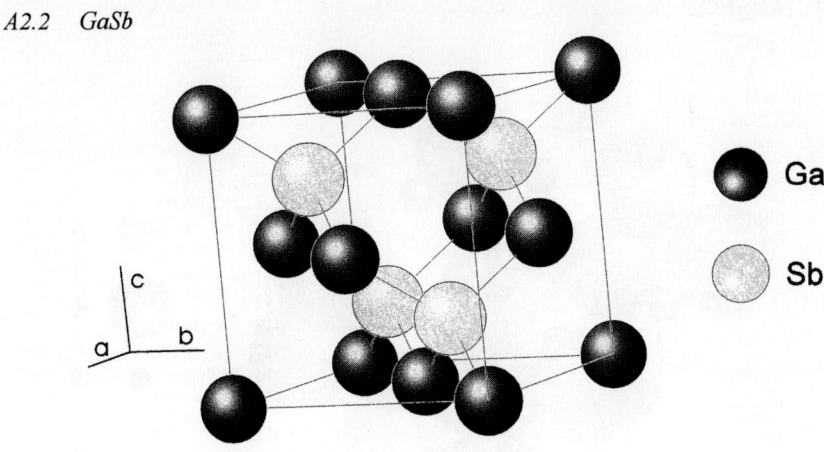

**Figure A2.2** *The GaSb structure (45).*

Atomic number / weight of components:    Ga: 31 / 69.723
                                         Sb: 51 / 121.75
Formulas per unit cell:          4
Crystal system:                  Cubic
Space group:                     $F\bar{4}3m$
Unit cell parameter:             a: 6.095Å
Atomic relative coordinates (47): table A2.2

**Table A2.2** *Atomic relative coordinates of GaSb .*

|     | x      | y      | z      |
| --- | ------ | ------ | ------ |
| Ga  | 0.0000 | 0.0000 | 0.0000 |
|     | 0.0000 | 0.5000 | 0.5000 |
|     | 0.5000 | 0.0000 | 0.5000 |
|     | 0.5000 | 0.5000 | 0.0000 |
| Sb  | 0.2500 | 0.2500 | 0.2500 |
|     | 0.7500 | 0.7500 | 0.2500 |
|     | 0.7500 | 0.2500 | 0.7500 |
|     | 0.2500 | 0.7500 | 0.7500 |

*A2.3   KH$_2$PO$_4$*

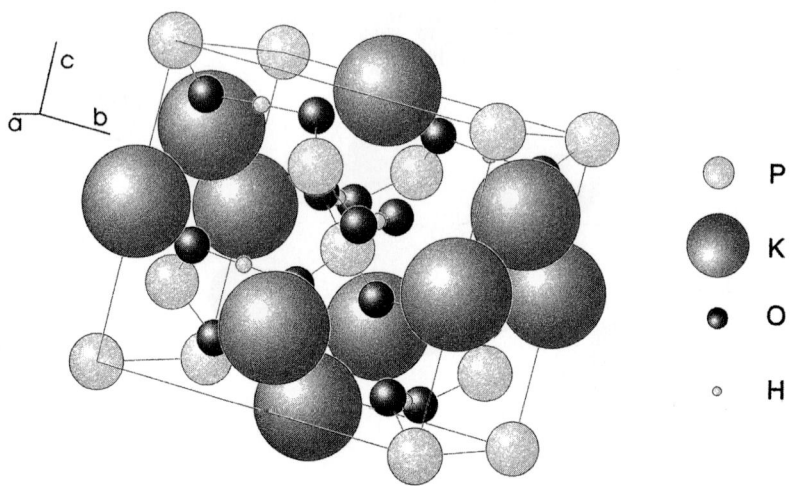

**Figure A2.3** *The  KH$_2$PO$_4$  structure (45).*

| Atomic number / weight of components: | P: 15 / 30.973762 |
|---|---|
| | K: 19 / 39.0983 |
| | O: 8 / 15.994 |
| | H: 1 / 1.00794 |

| | |
|---|---|
| Formulas per unit cell: | 4 |
| Crystal system: | Tetragonal |
| Space group: | $I\bar{4}2d$ |
| Unit cell parameters (48): | a: 7.453Å    c: 6.959Å |
| Atomic relative coordinates (49): | table A2.3 |

**Table A2.3** *Atomic relative coordinates of* $KH_2PO_4$.

|   | x | y | z |
|---|---|---|---|
| P | 0.0000 | 0.0000 | 0.0000 |
|   | 0.5000 | 0.0000 | 0.2500 |
|   | 0.5000 | 0.5000 | 0.5000 |
|   | 0.0000 | 0.5000 | 0.7500 |
| K | 0.0000 | 0.0000 | 0.5000 |
|   | 0.5000 | 0.0000 | 0.7500 |
|   | 0.5000 | 0.5000 | 0.0000 |
|   | 0.0000 | 0.5000 | 0.2500 |
| O | 0.0805 | 0.1440 | 0.1390 |
|   | 0.9195 | 0.8560 | 0.1390 |
|   | 0.8560 | 0.0805 | 0.8610 |
|   | 0.1440 | 0.9195 | 0.8610 |
|   | 0.4195 | 0.1440 | 0.1110 |
|   | 0.5805 | 0.8560 | 0.1110 |
|   | 0.6440 | 0.0805 | 0.3890 |
|   | 0.3560 | 0.9195 | 0.3890 |
|   | 0.5805 | 0.6440 | 0.6390 |
|   | 0.4195 | 0.3560 | 0.6390 |
|   | 0.3560 | 0.5805 | 0.3610 |
|   | 0.6440 | 0.4195 | 0.3610 |
|   | 0.9195 | 0.6440 | 0.6110 |
|   | 0.0805 | 0.3560 | 0.6110 |
|   | 0.1440 | 0.5805 | 0.8890 |
|   | 0.8560 | 0.4195 | 0.8890 |
| H | 0.2500 | 0.1440 | 0.1250 |
|   | 0.1440 | 0.7500 | 0.8750 |
|   | 0.7500 | 0.8560 | 0.1250 |
|   | 0.8560 | 0.2500 | 0.8750 |
|   | 0.7500 | 0.6440 | 0.6250 |
|   | 0.6440 | 0.2500 | 0.3750 |
|   | 0.2500 | 0.3560 | 0.6250 |
|   | 0.3560 | 0.7500 | 0.3750 |

*A2.4    KTiOPO₄*

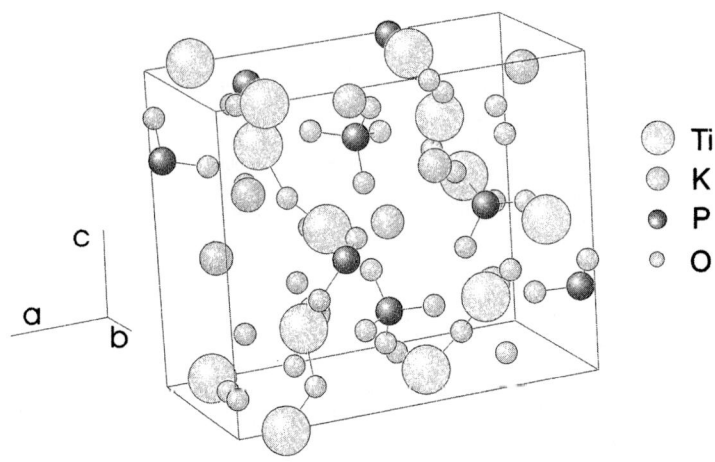

**Figure A2.4** *The KTiOPO₄ structure (45).*

| | |
|---|---|
| Atomic number / weight of components: | K: 19 / 39.0983 |
| | Ti: 22 / 47.88 |
| | O: 8 / 15.994 |
| | P: 15 / 30.973762 |
| Formulas per unit cell: | 8 |
| Crystal system: | Orthorhombic |
| Space group: | $Pna2_1$ |
| Unit cell parameters: | a: 12.819Å    b: 6.399Å    c: 10.584Å |
| Atomic relative coordinates (50): | table A2.4 |

**Table A2.4** *Atomic relative coordinates of KTiOPO₄.*

| | x | y | z | | x | y | z |
|---|---|---|---|---|---|---|---|
| K | 0.3781 | 0.7806 | 0.6880 | | 0.2525 | 0.5402 | 0.3718 |
| | 0.1053 | 0.6990 | 0.9332 | | 0.2528 | 0.4619 | 0.6008 |
| | 0.6219 | 0.2194 | 0.1880 | | 0.5141 | 0.5133 | 0.3497 |
| | 0.8947 | 0.3010 | 0.4332 | | 0.4897 | 0.5343 | 0.1170 |
| | 0.8781 | 0.7194 | 0.6880 | | 0.5996 | 0.8014 | 0.2208 |
| | 0.6053 | 0.8010 | 0.9332 | | 0.4066 | 0.8070 | 0.2589 |
| | 0.1219 | 0.2806 | 0.1880 | | 0.7248 | 0.5347 | 0.3561 |
| | 0.3947 | 0.1990 | 0.4332 | | 0.7768 | 0.9587 | 0.1097 |
| Ti | 0.3729 | 0.5001 | 0.9996 | | 0.8874 | 0.6894 | 0.9585 |
| | 0.2466 | 0.2695 | 0.7484 | | 0.8887 | 0.3082 | 0.0117 |
| | 0.6271 | 0.4999 | 0.4996 | | 0.7475 | 0.4598 | 0.8718 |
| | 0.7534 | 0.7305 | 0.2484 | | 0.7472 | 0.5381 | 0.1008 |
| | 0.8729 | 0.9999 | 0.9996 | | 0.9859 | 0.0133 | 0.8497 |
| | 0.7466 | 0.2305 | 0.7484 | | 0.0103 | 0.0343 | 0.6170 |
| | 0.1271 | 0.0001 | 0.4996 | | 0.9004 | 0.3014 | 0.7208 |
| | 0.2534 | 0.7695 | 0.2484 | | 0.0934 | 0.3070 | 0.7589 |
| P | 0.4981 | 0.3363 | 0.7397 | | 0.7752 | 0.0347 | 0.8561 |
| | 0.1808 | 0.5020 | 0.4872 | | 0.7232 | 0.4587 | 0.6097 |
| | 0.5019 | 0.6637 | 0.2397 | | 0.6126 | 0.1894 | 0.4585 |
| | 0.8192 | 0.4980 | 0.9872 | | 0.6113 | 0.8082 | 0.5117 |
| | 0.9981 | 0.1637 | 0.7397 | | 0.7525 | 0.9598 | 0.3718 |
| | 0.6808 | 0.9980 | 0.4872 | | 0.7528 | 0.0381 | 0.6008 |
| | 0.0019 | 0.8363 | 0.2397 | | 0.0141 | 0.9867 | 0.3497 |
| | 0.3192 | 0.0020 | 0.9872 | | 0.9897 | 0.9657 | 0.1170 |
| O | 0.4859 | 0.4867 | 0.8497 | | 0.0996 | 0.6986 | 0.2208 |
| | 0.5103 | 0.4657 | 0.6170 | | 0.9066 | 0.6930 | 0.2589 |
| | 0.4004 | 0.1986 | 0.7208 | | 0.2248 | 0.9653 | 0.3561 |
| | 0.5934 | 0.1930 | 0.7589 | | 0.2768 | 0.5413 | 0.1097 |
| | 0.2752 | 0.4653 | 0.8561 | | 0.3874 | 0.8106 | 0.9585 |
| | 0.2232 | 0.0413 | 0.6097 | | 0.3887 | 0.1918 | 0.0117 |
| | 0.1126 | 0.3106 | 0.4585 | | 0.2475 | 0.0402 | 0.8718 |
| | 0.1113 | 0.6918 | 0.5117 | | 0.2472 | 0.9619 | 0.1008 |

*A2.5    (CH$_2$NH$_2$COOH)$_3$H$_2$SO$_4$*

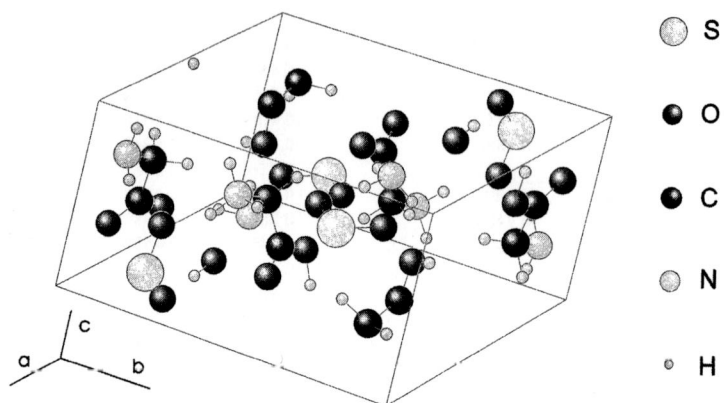

S

O

C

N

H

**Figure A2.5** *The (CH$_2$NH$_2$COOH)$_3$H$_2$SO$_4$ structure (45).*

| | | |
|---|---|---|
| Atomic number / weight of components: | S: 16 / 32.066 | |
| | O: 8 / 15.994 | |
| | C: 6 / 12.011 | |
| | N: 7 / 14.00674 | |
| | H: 1 / 1.00794 | |
| Formulas per unit cell: | 2 | |
| Crystal system: | Monoclinic | |
| Space group: | $P2_1$ | |
| Unit cell parameters: | a: 9.41Å    b: 12.64Å    c: 5.73Å | |
| | β: 110.4° | |
| Atomic relative coordinates (51): | table A2.5 | |

**Table A2.5** *Atomic relative coordinates of* $(CH_2NH_2COOH)_3H_2SO_4$.

| | x | y | z | | | x | y | z |
|---|---|---|---|---|---|---|---|---|
| S | 0.0002 | 0.2500 | 0.2282 | | | 0.6416 | 0.7092 | 0.8355 |
| | 0.9998 | 0.7500 | 0.7718 | | | 0.8954 | 0.0673 | 0.6983 |
| O | 0.8568 | 0.2472 | 0.0046 | | | 0.0808 | 0.9232 | 0.2912 |
| | 0.9652 | 0.2468 | 0.4579 | | H | 0.7070 | 0.2530 | 0.0490 |
| | 0.0836 | 0.1497 | 0.2064 | | | 0.2920 | 0.3300 | 0.8970 |
| | 0.0882 | 0.3397 | 0.2172 | | | 0.2630 | 0.1980 | 0.7770 |
| | 0.6086 | 0.2548 | 0.0777 | | | 0.4150 | 0.1430 | 0.1920 |
| | 0.4953 | 0.2653 | 0.6660 | | | 0.2520 | 0.1890 | 0.1780 |
| | 0.2202 | 0.5020 | 0.7774 | | | 0.4110 | 0.2600 | 0.3010 |
| | 0.4600 | 0.5298 | 0.7973 | | | 0.3270 | 0.5250 | 0.3270 |
| | 0.7848 | 0.4972 | 0.2405 | | | 0.3190 | 0.6450 | 0.4170 |
| | 0.5515 | 0.4808 | 0.2480 | | | 0.0640 | 0.4950 | 0.2950 |
| | 0.1432 | 0.7472 | 0.9954 | | | 0.0620 | 0.6000 | 0.1210 |
| | 0.0348 | 0.7468 | 0.5421 | | | 0.0560 | 0.6140 | 0.3850 |
| | 0.9164 | 0.6497 | 0.7936 | | | 0.5150 | 0.5010 | 0.0500 |
| | 0.9118 | 0.8397 | 0.7828 | | | 0.7160 | 0.4820 | 0.7310 |
| | 0.3914 | 0.7548 | 0.9223 | | | 0.7150 | 0.3550 | 0.6060 |
| | 0.5047 | 0.7653 | 0.3340 | | | 0.9600 | 0.5020 | 0.7110 |
| | 0.7798 | 0.0020 | 0.2226 | | | 0.9470 | 0.3940 | 0.8700 |
| | 0.5400 | 0.0298 | 0.2027 | | | 0.9500 | 0.3730 | 0.5890 |
| | 0.2152 | 0.9972 | 0.7595 | | | 0.2930 | 0.7530 | 0.9510 |
| | 0.4485 | 0.9808 | 0.7520 | | | 0.7080 | 0.8300 | 0.1030 |
| C | 0.4890 | 0.2571 | 0.8732 | | | 0.7370 | 0.6980 | 0.2230 |
| | 0.3409 | 0.2510 | 0.9182 | | | 0.5850 | 0.6430 | 0.8080 |
| | 0.3190 | 0.5302 | 0.6898 | | | 0.7480 | 0.6890 | 0.8220 |
| | 0.2683 | 0.5653 | 0.4166 | | | 0.5890 | 0.7600 | 0.6990 |
| | 0.6991 | 0.4727 | 0.3404 | | | 0.6730 | 0.0250 | 0.6730 |
| | 0.7485 | 0.4273 | 0.5982 | | | 0.6810 | 0.1450 | 0.5830 |
| | 0.5110 | 0.7571 | 0.1268 | | | 0.9360 | 0.9950 | 0.7050 |
| | 0.6591 | 0.7510 | 0.0818 | | | 0.9380 | 0.1000 | 0.8790 |
| | 0.6810 | 0.0302 | 0.3102 | | | 0.9440 | 0.1140 | 0.6150 |
| | 0.7317 | 0.0653 | 0.5834 | | | 0.4850 | 0.0010 | 0.9500 |
| | 0.3009 | 0.9727 | 0.6596 | | | 0.2840 | 0.9820 | 0.2690 |
| | 0.2515 | 0.9273 | 0.4018 | | | 0.2850 | 0.8550 | 0.3940 |
| N | 0.3584 | 0.2092 | 0.1645 | | | 0.0400 | 0.0020 | 0.2890 |
| | 0.1046 | 0.5673 | 0.3017 | | | 0.0530 | 0.8940 | 0.1300 |
| | 0.9192 | 0.4232 | 0.7088 | | | 0.0500 | 0.8730 | 0.4110 |

*A2.6    ZnO*

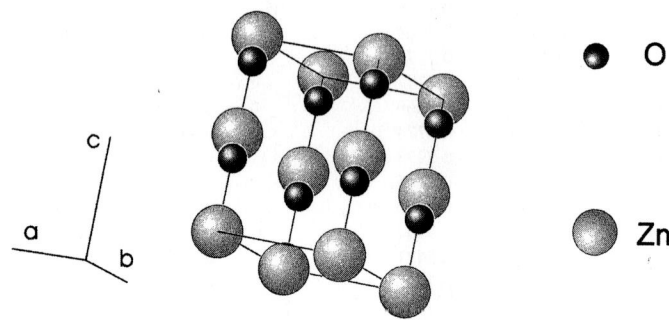

**Figure A2.6** *The ZnO structure (45).*

| | | |
|---|---|---|
| Atomic number / weight of components: | Zn: 30 / 65.39 | |
| | O: 8 / 15.994 | |
| Formulas per unit cell: | 2 | |
| Crystal system: | Hexagonal | |
| Space group: | $P6_3mc$ | |
| Unit cell parameters: | a: 3.249Å | c: 5.206Å |
| Debye temperature: | 370 °K | |
| Atomic relative coordinates (52): | table A2.6 | |

**Table A2.6** *Atomic relative coordinates of ZnO.*

| | x | y | z |
|---|---|---|---|
| O | 0.0000 | 0.0000 | 0.3825 |
| | 0.0000 | 0.0000 | 0.8825 |
| Zn | 0.0000 | 0.0000 | 0.0000 |
| | 0.0000 | 0.0000 | 0.5000 |

## A2.7 *LiNbO₃*

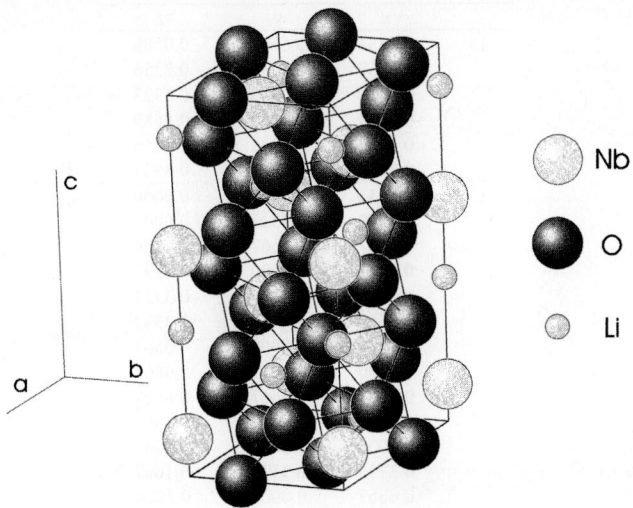

**Figure A2.7** *The LiNbO₃ structure (hexagonal unit cell) (45).*

| | |
|---|---|
| Atomic number / weight of components: | Li: 3 / 6.941 |
| | Nb: 41 / 92.90638 |
| | O: 8 / 15.994 |
| Formulas per unit cell: | 6 |
| Crystal system: | Rhombohedral |
| Space group: | *R3c* |
| Unit cell parameters: | a: 5.494Å   α: 55.867° |
| | hexagonal notation: |
| | a: 5.148Å   c: 13.863Å |
| Debye temperature: | 503 °K |
| Atomic relative coordinates (53): | table A2.7 |

**Table A2.7** *Atomic relative coordinates of LiNbO₃ (hexagonal unit cell).*

|      | x      | y      | z      |
|------|--------|--------|--------|
| Li   | 0.3333 | 0.6667 | 0.0589 |
|      | 0.6667 | 0.3333 | 0.2256 |
|      | 0.0000 | 0.0000 | 0.3923 |
|      | 0.3333 | 0.6667 | 0.5589 |
|      | 0.6667 | 0.3333 | 0.7256 |
|      | 0.0000 | 0.0000 | 0.8923 |
| O    | 0.3333 | 0.0000 | 0.0000 |
|      | 0.0000 | 0.3333 | 0.0000 |
|      | 0.6667 | 0.6667 | 0.0000 |
|      | 0.3333 | 0.0000 | 0.3333 |
|      | 0.0000 | 0.3333 | 0.3333 |
|      | 0.6667 | 0.6667 | 0.3333 |
|      | 0.3333 | 0.0000 | 0.6667 |
|      | 0.0000 | 0.3333 | 0.6667 |
|      | 0.6667 | 0.6667 | 0.6667 |
|      | 0.6667 | 0.0000 | 0.1667 |
|      | 0.0000 | 0.6667 | 0.1667 |
|      | 0.3333 | 0.3333 | 0.1667 |
|      | 0.6667 | 0.0000 | 0.5000 |
|      | 0.0000 | 0.6667 | 0.5000 |
|      | 0.3333 | 0.3333 | 0.5000 |
|      | 0.6667 | 0.0000 | 0.8333 |
|      | 0.0000 | 0.6667 | 0.8333 |
|      | 0.3333 | 0.3333 | 0.8333 |
| Nb   | 0.0000 | 0.0000 | 0.1037 |
|      | 0.3333 | 0.6667 | 0.2700 |
|      | 0.6667 | 0.3333 | 0.4366 |
|      | 0.0000 | 0.0000 | 0.6033 |
|      | 0.3333 | 0.6667 | 0.7700 |
|      | 0.6666 | 0.3333 | 0.9366 |

*A2.8    Na$_2$W$_4$O$_{13}$*

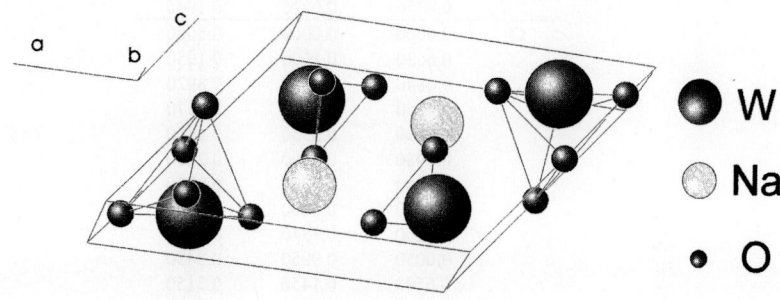

**Figure A2.8** *The Na$_2$W$_4$O$_{13}$ structure (45).*

| | |
|---|---|
| Atomic number / weight of components: | Na: 11 / 22.989768 |
| | W: 74 / 183.85 |
| | O: 8 / 15.994 |
| Formulas per unit cell: | 2 |
| Crystal system: | Triclinic |
| Space group: | $P\bar{1}$ |
| Unit cell parameters: | a: 11.163Å    b: 3.894Å    c: 8.255Å |
| | α: 90.60°    β: 131.36°    γ: 79.70° |
| Atomic relative coordinates (54): | table A2.8 |

**Table A2.8** *Atomic relative coordinates of* $Na_2W_4O_{13}$.

|     | x | y | z |
|-----|--------|--------|--------|
| Na  | 0.4300 | 0.2850 | 0.7360 |
|     | 0.5700 | 0.7150 | 0.2640 |
| W   | 0.1938 | 0.8382 | 0.2132 |
|     | 0.1744 | 0.9774 | 0.8054 |
|     | 0.8062 | 0.1618 | 0.7868 |
|     | 0.8256 | 0.0226 | 0.1946 |
| O   | 0.0000 | 0.0000 | 0.5000 |
|     | 0.6630 | 0.1600 | 0.1950 |
|     | 0.6840 | 0.1660 | 0.8920 |
|     | 0.9950 | 0.0050 | 0.1570 |
|     | 0.3460 | 0.8550 | 0.4850 |
|     | 0.1980 | 0.3890 | 0.2230 |
|     | 0.1480 | 0.4220 | 0.8270 |
|     | 0.3370 | 0.8400 | 0.8050 |
|     | 0.3160 | 0.8340 | 0.1080 |
|     | 0.0050 | 0.9950 | 0.8430 |
|     | 0.6540 | 0.1450 | 0.5150 |
|     | 0.8020 | 0.6110 | 0.7770 |
|     | 0.8520 | 0.5780 | 0.1730 |

# LIST OF SYMBOLS

$\omega$   circular frequency
$\varphi$   Debye function
$\mu$   linear absorption coefficient
$\nu$   linear frequency
$\delta$   path difference
$\phi$   phase difference
$\Theta_D$   Debye temperature
$\theta_B$   Bragg angle
$\Omega$   unit cell volume
$\lambda$   wavelength
$\Delta f$   atomic form factor correction
$a, b, c, \alpha, \beta, \gamma$   lattice parameters
$\mathbf{a}_1, \mathbf{a}_2, \mathbf{a}_3$   direct lattice vectors
$\mathbf{b}_1, \mathbf{b}_2, \mathbf{b}_3$   reciprocal lattice vectors
c   ligth velocity
d   interplanar distance
e   electron charge
f   atomic form factor
F   structure factor
h   Planck constant
(h k l)   crystallographic planes
I   intensity
$k_B$   Boltzmann constant
K, L, M   characteristic lines
m   atomic mass
$m_e$   electron mass
s   sample-to-detector distance
T   temperature
u   thermal oscillation amplitude
[u v w]   crystallographic direction
v   velocity
V   voltage
$x_n, y_n, z_n$   relative atomic coordinates
        of the n atom of the unit cell
$w_n$   weight fraction of component n
Z   atomic number

# REFERENCES

1. L. V. Azaroff, *Elements of X-Ray Crystallography* (McGraw-Hill, New York, 1968).

2. *International Tables for X-Ray Crystallography*, Vol. C (Kluwer Academic Publishers, Dordrecht, 1995), a) p. 177, b) p. 226.

3. S. S. Hasnain, J. R. Helliwell and H. Kamitsubo, *J. Synchrotron Rad.* **1** (1994) 1.

4. G. Margaritondo, *J. Synchrotron Rad.* **2** (1995) 148.

5. S. Suzuki, H. Kawata and M. Ando, *Photon Factory Activity Report* **VI** (1982/1983) 104.

6. W. Parrish, *Adv. X-ray Anal.* **8** (1996) 118.

7. H. E. Seemann, *Rev. of Sci. Instr.* **21** (1949) 314.

8. J. L. Amorós, *El Cristal* (Atlas, Madrid, 1990), ch. 1.

9. M. Senechal, *Crystalline Symmetries* (Adam Hilger, New York, 1990).

10. P. A. Doyle and P. S. Turner, *Acta Cryst.* **A24** (1968) 390.

11. D. T. Cromer and J. R. Waber, in *International Tables for X-Ray Crystallography*, Vol. IV (Kynoch Press, Birmingham, 1974) section 2.2.

12. W. L. Bragg, *The Crystalline State, Vol. I: A General Survey* (George Bell, London, 1933).

13. M. von Laue, *Ann. der Phys.* **41** (1913) 989.

14. D. C. Gillies, *J. Electrom. Mater.* **16** (1987) 151.

15. J. M. Bijvoet, N. H. Kolkmeyer and C. H. MacGillavry, *X-Ray Diffraction of Crystals* (Interscience Publishers Inc., New York, 1951), p. 36.

16. J. L. Amorós and M. Amorós, *Molecular Crystal: Their Transforms and Diffuse Scattering* (John Wiley and Sons, New York, 1968).

17. J. L. Amoros, M. J. Buerger, M. L. Canut, *The Laue Method* (Academic Press, New York, 1974).

18. J. R. Helliwell, J. Habash, D. W. J. Cruickshank, M. M. Harding, T. J. Greenhough, J. W. Campbell, I. J. Clifton, M. Elder, P. A. Machin, M. Papiz and S. Zurek, *J. Appl. Cryst.* **22** (1989) 483.

19. G. Christiansen and L. Gerwald, *J. Appl. Cryst.* **22** (1989) 397.

20. B. E. Warren, *X-Ray Diffraction* (Dover Publications, New York, 1990).

21. R. W. James, *The Crystalline State. Vol.II: The Optical Principles of the Diffraction of X-Rays* (George Bell, London 1948).

22. W. H. Zachariasen, *Theory of X-Ray Diffraction in Crystals* (Dover Publications Inc., New York, 1967).

23. B. D. Cullity, *Elements of X-Ray Diffraction* (Addison-Wesley, New York, 1978), a) p. 233, b) p. 504.

24. M. Canut-Amorós, *J.Comput. Phys.Commun.* **1** (1970) 293.

25. N. F. M. Henry, H. Lipson and W. A. Wooster, *The Interpretation of X-Ray Diffraction Photographs* (MacMillan & Co. Ltd., London, 1961), ch. 6.

26. B. Krahl-Urban, R. Butz and E. Preuss, *Acta Cryst.* **A29** (1973) 86.

27. B. Krahl-Urban, R. Butz and E. Preuss, *Laue Atlas* (Wiley, New York, 1974).

28. C. A. Cornelius, *Acta Cryst.* **A37** (1981) 430.

29. E. Preuss, *Comp. Phys. Commun.* **18** (1979) 261.

30. W. H. Huang, J. H. Christensen and R. J. Block, *Metall. Trans.* **2** (1971) 1367.

31. J. Lisbôa and D. F. Edwards, *Rev. Sci. Instrum.* **44** (1973) 1095.

32. R. A. Ploc, *J. Appl. Cryst.* **11** (1978) 713.

33. J. P. Riquet and R. Bonnet, *J. Appl. Cryst.* **12** (1979) 39.

34. J. Laugier and A. Filhol, *J. Appl. Cryst.* **16** (1983) 281.

35. C. Marín, A. Cintas and E. Diéguez, *J. Appl. Cryst.* **27** (1994) 846.

36. V. H. Hart and E. A. Rietman, *J. Appl. Phys.* **15** (1982) 126.

37. P. F. Fewster, *J. Appl. Cryst.* **17** (1984) 265.

38. J. S. Reid, *Comput. Phys. Commun.* **75** (1993) 259.

39. C. Marín and E. Diéguez, *J. Appl. Cryst.* **28** (1995) 839.

40. C. Marín and E. Diéguez, *J. Appl. Cryst.* **29** (1996) 198.

41. D. J. Winter, *Matrix Algebra* (MacMillan, New York, 1992).

42. G. Burns and A. M. Glazer, *Space Groups for Solid State Scientist* (Academic Press, Boston, 1990).

43. S. L. Chang, *Multiple Diffraction of X-Rays in Crystals* (Springer-Verlag, Berlin, 1984).

44. M. T. Santos, C. Marín and E. Diéguez, *J. Crystal Growth* **160** (1996) 283.

45. W. Kraus and G. Nolze, *POWDER CELL 1.8.* Federal Institute for Materials Research and Testing. Berlin, Germany.

46. S. F. Radaev and V. I. Simonov, *Sov. Phys. Crystallogr.* **37**(4) (1992) 484.

47. V. G. Giesecke and H. Pfister, *Acta Cryst.* **11** (1958) 369.

48. F. Jona and G. Shirane, *Ferroelectric Crystals* (Pergamon Press, Oxford, 1977).

49. B. C. Frazer and R. Pepinski, *Acta Cryst.* **6** (1953) 273.

50. I. Tordjman, R. Masse and J. C. Guitel, *Z. Kristallog.* **139** (1974) 103.

51. S. Hoshino, Y.Okaya and R. Pepinsky, *Physical Review* **115** (1959) 323.

52. S. C. Abrahams and J. L. Bernstein, *Acta Cryst.* **B25** (1969) 1233.

53. S. C. Abrahams, J. M. Reddy and J. L. Bernstein, *J. Phys. Chem. Solids* **27** (1966) 997.

54. K. Viswanathan, *J. Chem. Soc. Dalton Trans.* (1974) 2170.

# SUBJECT INDEX